A
SECOND GENESIS

Stepping-stones Towards the Intelligibility of Nature

A
SECOND GENESIS
Stepping-stones Towards the Intelligibility of Nature

Julian Chela-Flores

The Abdus Salam ICTP, Italy & Instituto de Estudios Avanzados,
República Bolivariana de Venezuela

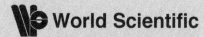 **World Scientific**

NEW JERSEY · LONDON · SINGAPORE · BEIJING · SHANGHAI · HONG KONG · TAIPEI · CHENNAI

Published by

World Scientific Publishing Co. Pte. Ltd.

5 Toh Tuck Link, Singapore 596224

USA office: 27 Warren Street, Suite 401-402, Hackensack, NJ 07601

UK office: 57 Shelton Street, Covent Garden, London WC2H 9HE

Library of Congress Cataloging-in-Publication Data
Chela Flores, Julián.
  A second genesis : stepping-stones towards the intelligibility of
nature / Julian Chela-Flores.
    p. cm.
  Includes bibliographical references and index.
  ISBN-13: 978-981-283-503-1 (hardcover : alk. paper)
  ISBN-10: 981-283-503-2 (hardcover : alk. paper)
  1. Exobiology. 2. Religion and science. 3. Philosophy of nature. I. Title.
  QH326.C48 2008
  576.8'39--dc22

                      2008035079

British Library Cataloguing-in-Publication Data
A catalogue record for this book is available from the British Library.

Printed in Singapore.

To the memory of my parents

Raimundo and Mercedes,

who first encouraged me to think beyond the frontiers of science.

Philosophy, as I shall understand the word, is something intermediate between theology and science. Like theology, it consists of speculations on matters as to which definite knowledge has, so far, been unascertainable; but like science, it appeals to human reason rather than authority, whether that of tradition or that of revelation.

Bertrand Russell[1]

Preface

We live in a golden age of the space, life and earth sciences. Never before did we have such detailed view of our cosmos, or understood in general terms its evolution, as well as the formation of our own planet. Never before have we been in a better position to approach the time-honored question:

Are we alone in the universe?

Yet, at the same time never before have we had such misrepresentation of the real frontiers of science. At a time when extreme specialization is necessary, due to the ever-increasing rate of discovery, it is necessary for scientists to explain the scope, limitations and successes of their research.

The general public has sometimes the mistaken impression that faith and reason are competing aspects of our culture, rather than complementary sectors of our humanity. To a large extent there is a dual aspect to keep in mind, firstly a vertiginous progress in science and technology, and secondly the complexity of human culture that goes beyond the frontiers of science.

The lack of appreciation of science by a large sector of our society may be due to the fact in science we do not always have a positive attitude to the question:

Does science need to be popularized?

The popularization of science assumes a central role in cultivating an appropriate relationship with the society that is funding our research. Valuable efforts to explain our work to a large sector of society have been made by scientists, journalists and even by gifted writers that, while not being practicing scientists, are nevertheless well informed of the main achievements of science. The successes that have received adequate coverage include the exploration of the Solar System, the discovery of other solar systems, and the ever-increasing accuracy with which we are already measuring the most basic physical phenomena that makes nature more intelligible.

Alas, such efforts in the popularization of science have not been sufficient. This is amply demonstrated by the present unnecessary, but sadly growing controversies generated by ignoring the natural frontiers of science. Controversies have also arisen by searching within the domain of the humanities what lies beyond their own frontiers. The experimental method that we have inherited from Galileo and his contemporaries lies inside science's frontier and defines its range of influence.

In spite of its evident difficulty, we have attempted to introduce the reader to a well-balanced review, in which science is inserted appropriately inside the wider domain of contemporary culture; this term is to be understood as the grand total of civilization's achievements in the millennia that have elapsed since the ancient paintings began to decorate the caves of Altamira and elsewhere, deep into prehistoric times.

In the past, books on Mesopotamian archaeology had attributed the invention of writing to a Sumerian living in Uruk about 3000 BC, although today the picture has been suggested to be somewhat more complex[2]. Even earlier, before 3100 BC, two important events took place. When a Giza race entered Egypt via the Delta, they brought new ideas with them including writing, thus introducing a second Middle-Eastern area into history[3].

Another major component of our culture emerged after the invention of writing, namely religion. Pharaoh Akhenaton introduced monotheism for a brief period (cf., Sec. 2.2). The permanent establishment of Abrahamic religions (Judeo-Christian-Islamic) followed the singular Egyptian event. The non-Abrahamic world creeds: Hinduism, Buddhism and Confucianism eventually completed the group of the main world religions.

The flourishing of culture received another major impact with the emergence of philosophy. Miletus, like other Ionian cities became a prosperous economic centre in the 7th and 6th centuries BC. The legendary Thales is considered to be the founder of philosophy. He assumed that water was the original substance, out of which all others are formed. In spite of the impression this statement makes on contemporary readers, it had the merit of stimulating thought and observation amongst his successors.

Finally, science, the last major component of culture began to take its present form after the seminal contributions of Copernicus and Galileo in the 16th and 17th centuries AD. Unfortunately, in spite of the flourishing of science after the Renaissance, there has been an ever-growing negligence of accepting the internal boundaries of each cultural sector. Philosophers have attempted to understand problems requiring the experimental approach of science. One example is consciousness. Science has also invaded domains beyond its well-defined experimental boundaries when, in spite of their academic training in science, rather than in the humanities, a restricted group of scientists in popularizing science have made statements (rather than theories supported by experiments) on questions for which training and familiarity with the humanities are essential. This attitude has conveyed the erroneous impression to the general public that believers should be exposed by what this restricted group of scientists regard as an interior misunderstanding on the part of the pious.

Sadly, overstepping the internal frontiers of culture has generated these controversies. At present ignoring these domains of culture has also taken place in a different context. The frontiers of science have been invaded with arguments that are foreign to the scientific tradition. The literal interpretation of the holy books of the monotheistic religions have led to argument, rather than to a constructive dialogue with science, especially with respect to the theory of evolution, which is not only one of the most remarkable scientific achievements, but it also represents the backbone of the life sciences. This is particularly relevant in the case of the Book of Genesis. Charles Darwin, by giving a theoretical foundation to natural history, improved considerably previous work in biology. Darwinism, like the whole of science, is clearly open to discussion, but this is an internal discussion that has to take place within the realm of science with publications in peer-reviewed journals.

We may find ways in the future of improving our rationalization of natural history. Fortunately, there are competent authors that have delineated clearly the internal frontiers of culture. The quotation at the beginning of this Preface by Bertrand Russell, the remarkable philosopher, mathematician and the 1950 Literature Nobel Laureate, is one such example. More recently, Professor John Cornwell has produced a timely, scholarly and clear presentation of these matters[4].

We hope that the present book will make a modest step in the correct direction for the communication of science, in order to dispel all anxious fears that either the humanities, or science, carry within themselves a certain danger for our deepest convictions in the realms of either faith, or reason. But such an undertaking is difficult, since academic training encourages expertise in restricted sectors of the cultural repertoire. To help the reader in this task, in Sec. 1.5 we make recommendations that will add to the great enjoyment of the present book by sharing the author's excitement for understanding the implications of the discovery of a second Genesis within our own lifetime.

Julian Chela-Flores,
Trieste, Italy,
October, 2008

Acknowledgments

The author has been very fortunate, and privileged, to have enjoyed the benefit of confronting his restricted training in biophysics with the views of exceptional scientists, philosophers and theologians.

Regular visits to Trieste, Italy, during the period 1968 till 1990 to the International Centre for Theoretical Physics provided the author with the unique opportunity of appreciating the remarkable depth in the work of Nobel Laureate Abdus Salam through many discussions in science, and its frontiers with the humanities. Fortunately, in the wider context of our cultural heritage Salam's book *Ideals and Realities* is accessible to everyone[1]. Salam shared with the author a passion for the problems related to the origin of life on Earth[2-4]. Salam continued his work in what we now call astrobiology till his untimely death, shortly after a volume was dedicated to him in his native Pakistan to commemorate his 70th birthday[5].

Many other scientists and humanists have contributed to the author's progress, He would like to take this opportunity to acknowledge especially his colleagues and collaborators Cyril Ponnamperuma[6,7], John Oro[8], Tobias Owen[9,10] and Francois Raulin[9,10-12]. Their advice, example and multiple discussions have helped him to find his own pathway along the frontiers of astrobiology.

The author has greatly benefited from the award of the 1998 UNESCO Chair of Philosophy that made possible the contact with the Venezuelan philosopher Ernesto Mayz Vallenilla[13]. The collaboration with the Spanish philosopher Roberto Aretxaga is acknowledged, especially for sharing a wide spectrum of common interests in numerous philosophical implications of astrobiology[14-18].

In addition, five events provided the author with additional sources of inspiration. They contributed to his approach of some difficult issues that are discussed in the following pages. Firstly, at very frontier of science and religion, the author acknowledges his participation in 1996 in the following two events:

Reflections on the birth of the Universe:
Science, Philosophy and Theology,

Evolutionary and Molecular Biology:
Scientific Perspectives on Divine Action.

These were unique meetings with corresponding publications. The first event was edited by Padre Eligio, Giulio Giorello, Gioachino Rigamonti and Elio Sindoni for the International School of Plasma Physics "Piero Caldirola"[19]. The second event was edited by Robert John Russell, William Stoeger S. J. and Francisco Jose Ayala for the Vatican Observatory and the Center of Theology and the Natural Sciences[20].

The author acknowledges the deeply inspiring sessions in 1998 of[21-22]:

Cattedra dei non-credenti[23].

This was an influential forum directed by His Eminence Cardinal Carlo Maria Martini from 1987 till he retired from the important See of Milan. Thirdly, the author would like to thank Charles L. Harper Jr. for his invitation to participate in the Templeton Foundation symposium held at the Harvard-Smithsonian Center for Astrophysics in 2003. The subject provided much inspiration for his work on the frontier of science. It centered on the question[24]:

Is the cosmos biocentric and fitted for life?

Finally, the interdisciplinary colloquium

The Origins: How, When and Where It All Started

was held in Rome in order to discuss the areas of astroparticle physics and astrobiology, two apparently different research lines that approach a common objective. Accademia Nazionale dei Lincei organized this event at the Centro Linceo Interdisciplinare "Beniamino Segre" in 2006. The author would like to thank Academicians. Francesco Bertola, Ernesto Carafoli, Giovanni Chieffi and Giancarlo Setti for encouraging him to reflect simultaneously on our cosmic and biologic origins, an activity that has provided him with deeper insights into the subject matter dealt with in this book.

The author would also like to thank astronomers Father George Coyne S. J.[25-28] and Frank Drake[29-31] for their thought-provoking contributions presented during the Trieste conferences that contributed to his additional insights in astrobiology.

Thanks are also due to Donald Goldsmith, Narendra Kumar for reading earlier versions of the manuscript. Their efforts, at earlier stages of the manuscript, helped the author to eliminate some preliminary inconsistencies and numerous difficulties. Special gratitude is expressed to Ms. Zhang Fang of World Scientific Publishers for her timely advice and enthusiasm that led to numerous improvements throughout the text.

We would like to acknowledge the kind permission of the Vatican Galleries to reproduce the magnificent painting of Raphael Sanzio that can be appreciated on he cover of the present work. Special thanks are due to Dr. Antonio Paolucci, Director of the Vatican Museums, who agreed to our requested use of an image of the Stanza della Segnatura (*L'Astronomia*). Dr. Daniela Valci from the Photographic Archives of the Vatican Museums, also provided valuable and timely assistance. At the same time we thank the courtesy of NASA for the 1995 Galileo mission image of the Jovian satellite Europa.

Finally, thanks are due to philosophers, scientists and theologians from different nations, and diverse backgrounds. Due to either their areas of expertise, or their areas of influence, his limited view across the "window of science" was enlarged to perceive nature through the windows of philosophy and theology. Their wise counseling, or active collaboration, allowed the completion of the present review of a very wide spectrum of subjects.

He would like to mention especially Afolabi Akindahunsi (biochemistry, Nigeria), Roberto Aretxaga (philosophy, Spain), Aranya Bhattacherjee (atomic physics, India), Suman Dudeja (chemistry, India), Giovanna Jerse (astronomy, Italy), Narendra Kumar (condensed matter physics, India), Mauro Messerotti (astronomy, Italy), Maria Eugenia Montenegro (paleontology, R. B. Venezuela), Nevio Pugliese (paleontology, Italy), Joseph Seckbach (microbiology, Israel), Vinod Tewari (micropaleontology, India) and Claudio Tuniz (paleoanthropology, Italy). And last, but not least, thanks are due to my wife, Sarah Catherine Dowling-Chela, for her permanent support of my work and for her guidance.

Contents

Chapter 1

Introduction

The question we wish to address concerns one of the most rewarding scientific and cultural activities that are possible at present. The frontier of science and the humanities is the exciting territory we begin to step on in the following pages. The fast development of technology has allowed us to address the issue of a possible "second Genesis", interpreted as the emergence of biological complexity elsewhere in the universe.

The reader is advised to refer especially to the following entries in the Glossary: Astrobiology, Biomarker, Biota, Galileo, Galileo mission, Newton, Prime Mover, Ptolemy, Raphael, Sagan, Second Genesis and Stanze della Segnatura.

1.1 An outline of the book

The question of intelligibility of nature has been a challenge in different fields of knowledge. One major aspect of the intelligibility of nature is the study of the origin, evolution, distribution and destiny of life in the universe, a scientific area that is known collectively as a single discipline: astrobiology.

Although this new science is far from reaching its maturity, it lies squarely within the frontiers of traditional research. Some of the deepest questions raised within astrobiology lie close to those raised within the humanities. From the point of view of philosophy and theology, it is conceivable to view conventional science as one aspect of a wider empiricism that would take into account such facts as the intelligibility of the universe. Some humanists feel that it is possible that a search for a rich empiricism could bring within human rationalization what lies beyond the scientific approach[1]. Since time immemorial humans have asked themselves whether we are alone in the universe on this "pale blue dot" (to borrow Carl Sagan's inspired description of the Earth[2]). Indeed, our planet can be conceived merely as a "dot", since it is circling around a common star that belongs to a common galaxy, in a universe that many scientists believe (right or wrong) to be one out of many universes. This is the central problem of astrobiology, namely to investigate whether life on Earth is but one example of a ubiquitous phenomenon.

However, metaphysics is the philosophical study to determine meaning, structure, and principles. Although this study is popularly conceived as referring to anything excessively subtle and highly theoretical, and although it has been the subject of many criticisms, it is presented by metaphysicians as the most fundamental and most comprehensive of inquiries, inasmuch as it is concerned with reality as a whole. Quite distinct in its nature, but equally profound is astrobiology[3]. Such an enquiry, which is the main objective of our research, is discussed at some length in this book.

Nevertheless, to paraphrase Sir Isaac Newton, we cannot avoid the feeling that our achievements in science are like having found small pebbles on an immense beach that represents knowledge. This is unavoidable since, in the tradition championed by Galileo, progress in science is a slow, serious and steady academic pursuit that encourages falsification by new experiments and repeatable observations. In other words, since Galileo science is strictly constrained by what can be verified by experiment, or careful repeatable observations, as in the case of astronomy. We feel deeply the need to encourage a dialogue between science and religion. Most of the pointless controversy that we are unfortunately experiencing at the present time can and should be avoided. Such an objective for a new book is possible, since science has well-defined frontiers within human culture. Similarly, faith does not have any need to justify itself in scientific terms, especially the most cherished values that religions have identified since the beginning of human reasoning. Yet, the depth of the questions in astrobiology should be the source of a fruitful dialogue with other sectors of the humanities, including theology.

1.2 The target audience of the book

We aim our writing at a level that will involve any reader who may be interested in the position of humans in nature, independent of whether the reader is more familiar with the humanities, or with science. In the process of writing such a book we had to combine different fields of knowledge, while retaining our objective of addressing a wide variety of educated people who, we hope, will find it engaging reading. The multidisciplinary nature of the question of intelligibility of nature forces upon us difficult areas that lie at the frontier of science and the humanities.

1.3 A second "Genesis"

The possibility of a second "Genesis", namely the emergence of life beyond the Earth, will inevitably present us with significant challenges arising from the ever-increasing pace of the exploration of the Solar System and, more generally, from the rapid progress of the space sciences. There is a compelling motivation for a careful discussion of the frontier of science and the humanities. Widespread general misunderstandings of the topics we have attempted to cover should be addressed seriously by practicing scientists. At present, even at the level of the school curricula much confusion arises due to a misconception of the real limits of science and the humanities. Indeed, science is an activity that is essentially experimental and supported by a reasonable theoretical background that is often presented in detailed mathematical language that is not easily understood by everyone. We have made a serious effort to avoid demanding from the reader detailed knowledge of science, philosophy, or theology. It is difficult for single individuals, either scientists, or humanists, to be able to approach reliably all aspects of contemporary culture. But it is nevertheless desirable to take additional modest steps in this challenging direction. Some experience in all aspects of culture is essential for a constructive, comprehensive and interdisciplinary discussion that dare to consider the implications of science in religion; for a mutual understanding, yet another book on the popularization of science is justified.

1.4 Three stepping-stones towards the intelligibility of nature

An old Chinese proverb states that a picture says more than a thousand words. The High-Renaissance-21st century image associated with this book summarizes adequately the main thrust of its subject matter. Raphael Sanzio (1509-1511) decorated the Room of the Signatura in the Palace of the Vatican. His painting of the "Prime Mover" is on a ceiling panel. The image represents the vision of the astronomers of the Middle Ages. The spirit of Ptolemaic cosmology and Christianity were being brought into harmony. Raphael aimed to reflect the interest of the Pope's library on the science of the early 16th century. The impact of the Copernican revolution was still ahead in the future, almost 30 years after the fresco of the Prime Mover had been completed. With Digges and Bruno a new cosmology was formulated and represented a first stepping-stone along the path towards the intelligibility of nature. We shall return to this topic in Sec. 5.5.

Charles Darwin placed a second stepping-stone firmly along the path towards the intelligibility of nature. In terms of evolution, it is comprehensible (in scientific terms) that a dolphin-sized brain of one of our ancestors (*Homo habilis*) should be able to grow in a geologic instant of time to the large sized brain of present-day humans[4], who are able to reflect on the question: *Are we alone in the universe?*

A Prime Mover is a concept introduced from Aristotelian cosmology in the Ptolemaic cosmos. He was capable of triggering the initial, eternal rotation at constant angular velocity of the outermost crystalline sphere of Heavens that lied even beyond the "Sixth Heaven", referred to in Dante's *Divina Commedia*[5] (cf., Secs. 2.3 and 5.2). The same poetic imagery may be used for explaining the possibility of searching for a second Genesis. Today we have to return to the Sixth Heaven (Jupiter's orbit), where the Prime Mover would be needed to call our attention to one Jovian satellite (Europa) that was going to be discovered by Galileo Galilei 100 years after Raphael completed his masterpiece. Europa's full view is shown on the book cover. It is underneath the Prime Mover. The image of the satellite was taken early in the Galileo mission. The center of the disk is near the Europan equator. Jupiter would be towards the left. The satellite's orbital motion is approximately away from the reader[6]. Europa is one of the targets of the current research of the author and of so many other scientists and engineers (cf., Chapter 12). The irrepressible excitement that this satellite generates on us is due to the possibility of there being a second Genesis underneath its icy surface, where we shall see later on, there may lie an ocean of liquid water (Sec. 12.4). The biota that we have conjectured to exist on Europa is only microbial, the lowest stage in the evolution towards intelligent behavior[7,8]. But instrumentation now at the stage of development[9,10] would be capable of detecting the "footprints of life" (biomarkers). We can conceive that within our lifetime we may lay a third stepping-stone firmly along the path towards the intelligibility of nature. The main themes of this book are firstly, to consider the possibility of identifying a second Genesis, either on Europa or elsewhere, and secondly, to review some of the implications on the humanities of such a phenomenal achievement.

1.5 Recommendations to the reader

Reading the book will require some effort. To facilitate this pleasant task we have provided a very lengthy, easily readable *Glossary and Short Biographies*. With over 200 entries, this part of the book forms the "backbone" of the text. We have made a special effort to introduce each chapter with an advice to read a few entries from the Glossary. This preliminary effort will reward the reader with a wider appreciation of the given chapter. The suggested entries will be an introduction, not only to especially relevant terms, or short biographical notes, but at the same time the reader will find that the Glossary is a rich repertoire of knowledge that is accessed by noticing that there are certain words highlighted in italics. In each entry these highlighted words refer to other significant entries in the Glossary. According to the time the reader devotes to this exercise, he will have the opportunity of gradually extending his appreciation of the various themes that are relevant for better appreciation of *A Second Genesis*. The detailed index with almost 800 entries can be used in conjunction with the study of the Glossary. The Index provides additional references to other sections of the book, in which certain terms are discussed. Another opportunity for a deeper understanding of the text lies in the detailed Bibliography. It contains about 400 publications. Most entries are original works, but many of them do not require excessive acquaintance with the specific cultural area that happens to be most familiar to a given reader.

Chapter 2

An Integrated Study of Western Civilization

To address interdisciplinary and intercultural issues, we use an integrated framework of the three branches of human culture: both aspects of philosophy, theoretical as well as moral philosophy (i.e., ethics), science and theology. Special attention has been paid to raising the right questions within their relevant cultural areas. In this Chapter we discuss topics related to the intelligibility of the universe. They include the origin of life[1], its evolution, distribution, destiny[2] and the Anthropic Principle[3]. These topics encourage us to dwell on implications of philosophy and theology.

The reader is advised to refer especially to the following entries in the Glossary: Aristotle, Compte, Darwin, Drake, Eukaryogenesis, Eukaryotic, Hoyle, Kant, Kenotic Process Theology, Logos, Monod, Naturalism, Oparin, Plato, Positivism, Russell, Ptolemy, Teleology, Thales and E. O. Wilson.

2.1 The three branches of culture

We shall first of all approach fine-tuning and the Anthropic Principle. Subsequently, we will discuss the intelligibility of life in the universe. Intelligibility is defined in terms of evolution of intelligent behavior in the cosmos. Whether the universe is fine-tuned for the emergence of intelligent life has been discussed in the past in a theological context. An approach within the natural boundaries of science is to study whether the evolution of intelligent behavior is inevitable, given what we know in the areas of the origin, evolution, and possible distribution of life in the universe. The assumption of the inevitability of the evolution of intelligence in the universe was already made implicitly several decades ago in the well-known project of Search for Extraterrestrial Intelligence (SETI).

The SETI project began to be developed in the early 1960s with the pioneering work of Frank Drake[4]. SETI is supported by solid technological development in order to scan systematically several windows of the electromagnetic spectrum. The main topic that we intend to discuss in the present chapter is the fitness of the cosmos for the origin and evolution of life and its relation with the Weak Anthropic Principle. First of all we begin with a discussion of various interrelations between the different branches of culture.

2.2 The growth of monotheism

Monotheism is the belief that there is one personal and transcendent God. Prayer and sacrifice may move a personal God. This concept is closely related with providence, which in turn is rooted in the belief in the existence of a benevolent, wise, and powerful deity or a number of beings that are benevolent and that are either fully divine or, at least, appreciably wiser and more powerful than man.

There was a common theme of the monotheistic religions in the world today. The early form of monotheism of the Egyptians lasted only a brief period during the 18th Dynasty under the influence of Amenophis IV, also known as Pharaoh Akhenaton (1353-1335 BC)[5]. Akhenaton devoted himself to the worship of the Aton, erasing all images and all writings containing the word "gods". Sadly, Atenism was rejected by the counter-revolution of Horemheb (1335-1305 BC). Akhenaton's alteration of the religious life of Egypt was profound. He laid the greatest stress on his own divinity. He declared the earliest form of Incarnation, namely a manifestation of the single God whose name was Aton, who represented the Sun's disk, unlike earlier representations of God during the Renaissance in the beautiful paintings of Raphael in the Vatican and Tintoretto in the Accademia (Venice)[6].

These events took place almost half a millennium before King David of Israel. For the present-day monotheistic religions Judaism, Christianity and Islam, creation was rationalized in terms of nature that was assumed to be subject to regular phenomena. This tendency eliminated the mythological creation of the world that had been based on many deities, such as those represented in the Greek pantheon.

Creation of the universe and life is intelligible to all monotheists as the three manifestations of monotheism read the same Book of Genesis. In monotheism an impersonal rational principle, a logos or nous, dominated the development of philosophical thought. These philosophical concepts were imported into a theological context in Western civilization, from the writings of Plato and, later, of Aristotle. The acceptance of theism or, the belief in a single divine being, a single creator, known with various names: Aton, Yahweh, the Word and Allah was a gradual process that spanned the temporal period of two millennia, from 1350 BC in Egypt till the 7th century AD in Arabia.

Atenism removed the complete Egyptian pantheon, anticipating King Solomon at the time of the construction of the First Temple in 970 BC. After the return from the Exile and the laying out of the foundations of the Second Temple by Zerubbabel in 520 BC, all deities other that the single God were removed. Atenism differed from subsequent forms of monotheism in the role played by the Head of the religion; while Christianity, more specifically, Catholicism, declares the infallibility of the Head of the Church, Atenism went beyond that in declaring that the founder of the religion was divine himself. Monotheism went beyond the restriction of an ethnic religion when Christianity spread throughout the Roman Empire, mainly due to Paul of Tarsus (died 65 AD): St. Paul had an entirely different conception of Christianity from the disciples of Jesus of Nazareth. Jew and gentile were to be considered as equals in the pursuit of monotheism. Finally, tradition regards Islam as the final evolution of monotheism[7].

The rapid spread of Islam went beyond the boundaries of what had once been the Roman Empire. In the year 732 AD the battle of Poitiers in France finally defined the rapid spread of Islam beyond the confines of what was the Roman Empire, virtually from the Atlantic coast of the Iberian Peninsula to the river Ganges in the Indian subcontinent.

2.3 The emergence of science and philosophy

Ever since ancient times the human intellect focused its attention on the study of nature and its regularities, rather than on arbitrary cosmogonies. The bases of these searches led to the origin of a cultural activity that gradually crystallized into what we now recognize as science. Good examples illustrating the influence of the evolution of science in the areas that were under the influence of monotheism is the development of astronomy in the Arab world[8] and the evolution of one aspect of cosmology (cosmogony).

There is a consensus, dating back at least to the 4th century BC and continuing to the present, that the first Greek philosopher was Thales of Miletus, who flourished in the first half of the 6th century BC[9]. Thales tried to give the mathematical knowledge that he derived from the Babylonians a more exact foundation and by using it for the solution of practical problems-such as the determination of the distance of a ship as seen from the shore or of the height of the Pyramids. Though he was also credited with predicting an eclipse of the Sun, it is likely that he merely gave a natural explanation of one on the basis of Babylonian astronomical knowledge. Thales was considered the first Greek philosopher because he was the first to give a purely natural explanation of the origin of the world, free from all mythological ingredients. He upheld that everything had come out of water-an explanation based on the discovery of fossil sea animals far inland. His tendency (and that of his immediate successors) to give non-mythological explanations of the origin of the world was undoubtedly prompted by the fact that all of them lived on the coast of Asia Minor surrounded by a number of nations whose civilizations were much farther advanced than that of the Greeks and whose mythological explanations differed greatly both among themselves and from those of the Greeks.

It appeared necessary; therefore, to make a fresh start on the basis of what a person could observe and figure out by looking at the world as it presented itself. This procedure naturally resulted in a tendency to make sweeping generalizations on the basis of rather restricted but carefully checked observations.

Anaximander (610-545 BC, circa) was a Greek philosopher often called the founder of astronomy, the first thinker to develop a cosmology, or systematic philosophical view of the world. Anaximander may be considered to be at the basis of the Western concept of the universe. He is thought to have been a pupil of Thales of Miletus. His theories departed from earlier, more mystical conceptions of the universe and anticipating the achievements of later astronomers. A basic contribution in Anaximander's theory was his rejection of that the Earth was somehow supported in the heavens; instead, he placed the Earth at the centre of the universe—an essential element inserted 600 years later by Ptolemy. Anaximander's assumption is one of the most long-lived assumptions in science for it lasted about two thousand years, before the proposal and general acceptance of the heliocentric theory.

Indeed, Ptolemy was active in Alexandria from AD 127-145. He was the most influential astronomer of the first millennium. He considered the Earth to be at the centre of the universe (the "Ptolemaic system"). Very little is known of his life. His main contribution to astronomy was the "Almagest". Ptolemy extended some of the work of Hipparchus, who had compiled a star catalogue containing 850 stars (cf., Sec. 5.2). Ptolemy's catalogue included 1022 stars. The Ptolemaic system is introduced in the first section of the Almagest. Ptolemy rather complex geometrical explanation of the retrograde motion of planets with the Earth immovable at the centre of the cosmos was very successful in the sense that it prevailed in Western culture for over a millennium.

Ptolemy understood that planets were closer to the Earth than the "fixed" stars, but he introduced in his system a concept that not even Copernicus discarded. Ptolemy believed in the physical existence of crystalline spheres on which the heavenly bodies were supposed to be attached. We had to wait for the contributions of Digges and Bruno to reach the present-day view of the cosmos (cf., Chapter 5, Sec. 5.4), in which the stars spread isotropically and homogeneously over the universe.

Ethics is that aspect of philosophy that is concerned with moral philosophy. In fact, the only frontier that is of some interest from the point of view of an integrated discussion of all of human culture is whether ethics and science can be integrated. The idea of a naturalistic ethics arose from a teleological outlook.

Naturalism relates scientific method to philosophy by maintaining that all beings and events in the universe are natural. Consequently, all knowledge of the universe falls within the domain of scientific investigation. Naturalism makes allowance for effects that are not currently within the range of scientific enquiry, provided that knowledge of it can be retrieved indirectly. In other words, natural objects could conceivably be under the influence of some effects beyond the present range of physics and chemistry in a measurable way[10]. Naturalism assumes that nature is in principle comprehensible. In nature there is both regularity and unity implying objective laws. Man's search for proofs of his beliefs is seen as a confirmation of naturalistic methodology. Naturalists point out that even when one scientific theory is rejected and replaced by another, this only means that theories change; but the methodology remains.

The idea of a naturalistic ethics arose in the context of Aristotelian philosophy[11]. In this aspect of Greek philosophical discussions there is inherent in each natural kind of thing an appropriate manner in which the given kind behaves. In Aristotelian philosophy the purpose of life, for instance is, in the case of man, to achieve in an objective sense what is appropriate to him. Clearly, at this early stage in the development of philosophical thought in Western civilization it was somewhat difficult to escape from teleological arguments. Darwinism has taught us that we are to avoid such arguments. But this aspect of Darwinism still needs to be inserted comprehensively into ethics in a satisfactory manner[12].

Philosophy of nature is the subject concerned with investigating issues regarding the actual features of nature as a reality. The main contribution of the philosophy of biology is to separate linguistic difficulties from matters of substance. From the point of view of this book, in which we explore the fitness of the universe for the origin and evolution of life, the main questions that concern us are related to the philosophy of biology.

The Lamarckian view of evolution involves the inheritance of acquired characteristics. Such an approach claims that evolution involves a deterministic finalism, or directedness toward an end. Several positions have been adopted in the development of evolution (cf., Sec. 12.5).

Firstly, natural selection is a non-random element in evolution. Secondly, it has been argued that chance in mutation and selection, in addition, the unpredictability of environmental change, militates against the formulation of deterministic laws. These two positions would support the view that the course of evolution can never be predicted. In spite of this there is ample evidence that convergence in the space sciences and the life sciences present us an alternative that will be discussed in this book in Chapters 7 and 8. It is not excluded that general predictions would be possible in spite of the contingency of evolution. However, to a certain extent the outcome of the evolutionary process can be foreseen to a large extent.

This point has been documented[13-15]. Another aspect of the frontier of science and humanities is the philosophy of science. First of all this subject attempts to elucidate the elements involved in scientific enquiry and thus becomes a topic for explicit analysis just as are the subdivisions of philosophy.

2.4 Faith and reason

An answer to the fundamental question of the relation man and universe requires that we keep in mind the manner in which various elements of culture evolved within their domains of competence. The most significant questions were raised during the 17th and 18th century in Europe: the Enlightenment was an intellectual movement in which truth was to be approached through reason[16]. Reason is to be understood as the process through which human intelligence is applied to religious traditions and to revelation. Ideas concerning God, reason, nature, and man went into a synthesis that had many supporters. The most distinguished thinkers of this period were, amongst others, Descartes, Diderot, Montesquieu, Pascal, Rousseau, Voltaire and Kant. This movement was influential on the development of ethics, philosophy, science and theology. Besides, the Enlightenment also stimulated the growth of art, politics and history. Reason, instead of increasing knowledge, was the main theme underlying most innovations of this period. The thinkers behind this movement searched for a deeper understanding of the cosmos. Rationalists strived towards more freedom, knowledge and happiness.

Isaac Newton's theory of gravitation was formulated in the language of mathematics and was considered to be of universal validity[17]. This eliminated the bases for geocentrism. The equivalent removal of anthropocentrism had to wait for the Darwinian theory of evolution.

Reason became predominant. Increasing the available knowledge took a second place in the priorities of this 18th century movement. This was a period in which philosophy flourished. The major force behind the Enlightenment was the widespread acceptance of the Newtonian theory with its basis solidly lying on observation and experience. The assumed universality of Newtonian theory influenced areas of knowledge, ranging from physics to religion and essentially to all forms of knowledge.

Voltaire's real name was Francois Arouet (1694-1778). He was influenced by the philosophical empiricism of Locke. His "Dictionnaire philosophique" gathered his views on metaphysics, religion and ethics[18]. He advocated some form of deism. (God is not involved in the world in a personal way, but instead He is responsible for its laws.) Voltaire was critical of the contemporary clergy. Although he defended the new aspects of the Enlightenment, he did not ignore the limitations of reason itself.

Voltaire touched upon arguments that even today concern the debate of faith and reason. He believed in the argument of design. In other words, he held the view that the teleological argument is supported from an observation of the general laws of the universe, which are to be interpreted as revealing that the world functions toward ends or goals. He was an immediate predecessor of greater philosophers, such as David Hume and Immanuel Kant. The German philosopher Kant (1724-

1804) distinguished himself in the theory of knowledge, ethics, and aesthetics. By so doing he influenced all subsequent philosophy (the "post-Kantian" period). Kant was the leading philosopher of the Enlightenment and one of the greatest philosophers of all time, sometimes equated with Plato and St. Augustine, as the greatest thinkers. In him were subsumed new trends that had begun with Rationalism (stressing reason) of René Descartes and Empiricism (stressing experience) of Francis Bacon.

2.5 The Enlightenment

Voltaire accepted the classical ideal of the universal morality of man. In fact, the Enlightenment produced the first modern theory of ethics. Under the influence of the Enlightenment the French philosopher Auguste Comte (1798-1857) founded a movement advocating that intellectual activities should be confined to observable facts. The reason why this movement was called "positivism" is that Comte called observable facts "positive". Positivism can be considered as a philosophical system of thought maintaining that the goal of knowledge is simply to describe a phenomenon that we experience, not to question whether it exists or not. This point of view was developed much later by a group of philosophers working in Vienna in the 1920s and 1930s. They were known as the "Vienna Circle". The nucleation of the group began with Moritz Schlick, when he settled in Vienna in 1922[19]. Some of the Vienna Circle's members were Rudolf Carnap, Hans Reichenbach, founder of the Berlin Circle, Herbert Feigl, Philipp Frank and Otto Neurath. Karl R. Popper and H. Kelsen were related with the Vienna Circle, although strictly they did not belong to it.

This group of philosophers maintained that scientific knowledge is the only kind of factual knowledge. The distinctive aspect of this version of extreme positivism was an attempt, referred to as "logical positivism", to develop knowledge based on experience (empiricism), with the help of mathematics and logic. They insisted on the soundness of logical analysis of scientific knowledge. Indeed, logical positivists appealed to the earlier contributions of Russell and Wittgenstein.

The Vienna Circle, to which we shall return in the next section, maintained that all traditional doctrines are to be rejected as meaningless. They were generally hostile to metaphysics and theology. The Vienna Circle went beyond positivism in maintaining that the ultimate basis of all knowledge rests on experiment. Although scientists have adopted this philosophy, either consciously or unconsciously, modern science begins with Galileo Galilei, who initiated the tradition of formulating theories based on observation and experiments. No underlying philosophy was adopted then, or need to be adopted now, beyond the dialogue theory/experiment. There are a large number of issues that science cannot handle or even formulate (cf., Sec. 13.8). In his *History of Western Philosophy* Russell makes this point[20]:

> *Almost all the questions of most interest to speculative minds are such as science cannot answer.*

Positivism avoided all considerations of ultimate issues, including those of metaphysics and religion. However, as anticipated by Russell, the reduction of all knowledge to science is a matter that debate has not yet settled. Reasoning from first and final causes are precisely topics relevant to the subject matter of this book. To 19th century scientists, the problems of the origin and distribution of life in the universe were issues that were to be excluded from the scientific discourse. We have attempted to show that these problems are approachable by scientific methods.

These subjects have a consistent history of valuable efforts by some of the best scientific minds of the recent past. The long list of such scientists began with two distinguished scientists Alexander Oparin (1894-1980) and John Burdon Sanderson Haldane (1892-1964). A long line of organic chemists and life scientists followed them.

Firstly, Oparin was an unusual scholar, close to what sometimes is called a "renaissance man". He faced a problem that was basically philosophical in nature, but brought to bear upon it history as well as science. Oparin referred both to the philosophy of Aristotle, as well as to later writings of Saint Augustine. In spite of his biochemical training, Oparin was familiar with various fields from astronomy to chemistry, from geology to biology, form philosophy to theology. This broad approach to science and its implications to humanities led him to significant contributions to the early development of the origin of life studies. Secondly, Haldane, was a pioneer like Oparin in the studies of the origin of life on Earth. Haldane was a British biologist. He was a very influential scientist. He studied the relationships of Mendelian genetics on evolutionary theory, enzymology and genetics, and theoretical biology.

Extreme aspects of first causes and ultimate ends are naturally inserted in the studies of the origin of life; yet, neither of the two problems (origins and distribution of life) is solidly set on scientific bases: it has been impossible to synthesize a living organism so far, and no signal from an extraterrestrial civilization has yet made contact with the highly sensitive radio telescopes. In view of this unsettled state of affairs, it seems reasonable that the deepest questions in astrobiology should be of interest to the humanities.

2.6 The unity of nature

Both science and religion are concerned with the common understanding of life in the universe. Since they both address deeper issues concerning origins and destiny, the convergence of science and religion is an important topic to which we shall return in this chapter. With subsequent progress in philosophy, science and theology, convergence seems unavoidable. There does not seem to be any evident convergence in culture at present. The status of the relationship between these three disciplines is of considerable interest.

> *Contemporary culture demands a constant effort of synthesis of knowledge and of an integration of our understanding... but if the specialization is not balanced by an effort aiming to pay attention to relationships in our understanding, there is a great risk of arriving at a "splintered culture", which would in fact be the negation of the true culture*[21].

We should aim to avoid a splintered culture in the present integrated approach to the questions regarding the fitness of the universe for the origin and evolution of life. The study of the implications of the search for extraterrestrial life would be a less difficult task if our culture were better integrated. Unfortunately, we are still far from having a unified culture. It has been mentioned that culture runs the risk of being "splintered", if specialization is not balanced by an effort to pay attention to relationships between different aspects of our understanding of nature. We have attempted to insert the origin of life in the context of the origin of the universe itself[22].

Our aim is to illustrate some relationships between the origin of life and other aspects of culture, namely, theology, philosophy and art. Although we are still not anywhere near unifying all learning, such "Ionian enchantment" is for some still desirable[23]. Indeed, the Greek philosophers in Ionia pursued this objective in ancient times. In the early 20th century Ionian enchantment was the driving force of the Vienna Circle of Logical Positivists (cf., Sec. 2.5).

Logical positivism intended to exclude from the realm of the cognitively meaningful those statements for which supporting or refuting evidence (satisfactory to scientists and philosophers) can be found. This clearly led to a confrontation with natural theologians. Independent of the shortcomings of the Viennese positivists[24], discussing a given problem in a wider framework than that of a specialized topic, has today a practical advantage for us: this wider framework will allow access to a richer array of metaphors, examples and analogies.

The unification of culture has its basis in the belief that a common body of inherent principles underlies the entire human endeavor. The Enlightenment thinkers of the 17th and 18th had this belief. They assumed a world in which knowledge is unified across the sciences and the humanities. Edward O. Wilson calls this common groundwork of explanation that crosses all the great branches of learning "consilience". He maintains that we can indeed explain everything in the world through an understanding of a handful of natural laws. In this view the world is a material world that is organized by laws of physics and evolves according to the laws of evolution.

Against this tendency we find cultural relativism. (In Sec. 12.1 we shall return to this concept.) In examining how a few underlying physical principles can explain everything from the birth of stars to the workings of social institutions to art, consilience offers fresh insight into what it means to be human. This approach, in turn, will facilitate our discussion of the all-important problem of the origin, evolution and distribution of life in the universe. For simplicity, in Chapter 1 we referred to this study as "astrobiology". More specifically, we will focus on a deep question:

> *Are we alone in the universe and, if not, what is the place of Man in the universe?*

We wish to put in evidence changes that the current view of astrobiology brings onto different aspects of culture, namely, literature, art, theology, philosophy and science. We begin our work by recalling how culture itself emerged gradually in a series of steps.

2.7 The evolution of culture

The earliest records of cultural activity in humans correspond to works of art. The Magdalenian "culture" left some fine works of primitive art. This group of human beings flourished from about 20,000 to 11,000 years before the present (yr BP). For instance, we can refer to the 20,000 year old paintings on the walls of the cave discovered in December 1994 by Jean-Marie Chauvet in Southeastern France. Indeed, the birth of art in the Magdalenians' caves, is one of the most striking additions to the output of humans that entitle us to refer to the groups that produced these fine works, as cultures, rather than industries, a term that is reserved to the group of humans that produced characteristic tools, rather than works of art.

On the other hand, as briefly mentioned in the Preface, traditionally the earliest examples of a writing system have been accepted to be those found in the Sumerian settlements of the Fertile Crescent some 5500 yr BP. Independently writing was introduced in the Nile Valley during the Pre-Dynastic Period. Recent discoveries suggest that the Egyptian system may have anticipated the Sumerian. Besides the great achievements of the Ancient Kingdom, the Sphinx and the pyramids, a remarkable literature was produced on the shores of the Nile, which has been largely translated during the 19th century. These compositions are narratives, tales, teachings and poetry. An example form the Middle Kingdom (most probably the 12th Dynasty, 1991-1976 BC) is a substantial body of religious literature, including the *Cycle of Songs in Honor of Sesostris III*[25].

The Egyptian civilization led to another relevant step forward that was later to influence Western civilization. We have seen earlier that the most ancient statements on monotheism may be traced back to the New Kingdom (1379-1362 BC). It is still debatable whether the system initiated during the 18th Dynasty may be called monotheism. However, theology is a discourse about God, or the science that treats the divine; the earliest such discourses that influenced directly Western civilization may only be traced back to the oral traditions of ancient Israelites, which were incorporated into the main body of the Torah. It is in theological arguments that we find the earliest records of discussions of Man's place in the universe, a topic that is relevant to astrobiology. In the ancient texts quoted above we find early references to the origin of the earth and the origin of life.

Philosophers came next in the evolution of culture. Its origin may be traced back to Greek philosophy, but their interest in the existence of Man may best be illustrated with the preliminary efforts of Rene Descartes (1596-1650, cf., Sec. 2.4), who is usually considered to be the founder of modern philosophy. Descartes is probably the first man of high philosophic capacity whose outlook is profoundly affected by the new physics and astronomy[26]. Descartes' arguments illustrate some points relevant to the present context. He regarded the bodies of men and animals as machines; in particular, he regarded animals as automata, governed entirely by the laws of physics. He assumed that men were intrinsically different from animals in possessing a soul. Descartes attempted to describe how soul and body interacted, not surprisingly without any real success. To arrive at a continuous argument on the origin of Man, science had to await the arrival of Charles Darwin and his contemporaries.

As opposed to the incorporation of philosophers in Western civilization, scientists are relative newcomers to the question of the origin and evolution of life. Darwinism did not address the question of the origin of Man in the universe in the Cartesian dualistic way (body/soul), but restricted itself only to the observable attributes of Man. Since the time of Galileo the scientific approach differs from that of theologians, in the sense that no appeal to revelation is introduced into our discussions. Scientists also differ from philosophers since no exclusive appeal to rational discussion is expected.

To sum up, the evolution of culture, first through the first appearance of art, writing and literature, eventually led to those aspects of culture in terms of which reasonable questions on the origin and destiny of Man could be formulated.

Theologians are mainly concerned with the metaphysical concepts of "spirit" and "soul", which do not appear in the scientific literature; one could say that these concepts could not appear, since they are not subject to experimental reproducible tests. Instead, the existence of spirit and soul are based on revelation in the traditional teachings of the monotheistic religions.

2.8 Is the position of humans special in the tree of life?

Some biologists have favored a different point of view from those radio astronomers that have been searching for extraterrestrial intelligence[27]. The reasons behind this dichotomy may be illustrated with a few facts drawn from taxonomy. From the perspective of biology, human beings represent only a single species among four thousand mammals. Yet, this is a small number when compared with the 30 million species that are expected to constitute the whole of the Earth biota. One aspect of this bewildering abundance of species, of which humans are only one, has led to the metaphor:

> *If the history of evolution were to be repeated, such an alternative world would teem with myriad forms of life, but certainly not with humans*[28].

Our main concern is not the origin and evolution of our own species. Our main concern is rather the likelihood that the main attributes of Man would raise again, if the history of evolution starts all over again elsewhere, rather than on a hypothetical Earth. We are mainly concerned with the repetition of biological evolution in a planet, or satellite, that may have had all the environmental conditions appropriate for life. Such attributes are, for example, a large brain and consciousness. These features of Man evolved from primates over the last 5 to 6 million years. This is in plain contrast with mollusks. Members of this large phylum of invertebrates, to which snails, mussels and squids belong, have survived since the Lower Cambrian. Their first appearance occurred 500 million years ago, about 100 times earlier than the first appearance of Man.

2.9 What is needed to duplicate our intelligence?

In an extraterrestrial environment the evolutionary steps that led to human beings would probably never repeat themselves. However, the possibility remains that a human level of intelligence may be favored when a combined effect of natural selection and cultural evolution are taken together. This is independent of the particular details of the tree of life that may lead to the intelligent (non-human) organism. To put it briefly, the role of contingency in evolution has little bearing on the emergence of a particular biological property[29]. We can illustrate the inevitability of the emergence of particular biological properties with an example of convergent evolution, a phenomenon that is well known to students of evolution.

For example, in the phylum of mollusks the shells of both the camaenid snail from the Philippines, or a helminthoglyptid snail from Central America, resemble the members of European helcid snails[30]. These distant species (they are grouped in different Families), in spite of having quite different internal anatomies, have grown to resemble each other outwardly over generations of response to their environment. In spite of considerable anatomical diversity, mollusks from these distant families have tended to resemble in a particular biological property, namely, their external calcareous shell.

At this stage the question of whether our intelligence is unrepeatable goes beyond biology and the geological factors mentioned in the metaphor on the repetition of the history of evolution. Indeed, the question is rather one in the domain of the space sciences; in particular, the question of whether we are alone in the cosmos concerns astrometry measurements for the search for extra-solar planets and space-bound probes that would search for biosignatures. This activity has led to the current revolutionary view that solar planets are not unique environments that may be conducive to the origin, evolution and distribution of life in the universe[31].

The presence of numerous planets in the cosmic neighborhood of the Sun, argues in favor of the plausibility of life in the universe. It seems likely that if the right environments exist elsewhere in our own galaxy, some of the biological attributes of Man may have repeated themselves.

2.10 Constraints imposed by theology on our view of life

A separate question concerns the search for microbial life in our own solar system. So we believe it is more appropriate to shift our attention away from "attempting a full and coherent account of the phenomenon of Man"[32]. Instead, we feel that progress of molecular biology forces upon us a search for a full and coherent account of the first transcendental transition in cellular evolution, which may be referred to as the phenomenon of the eukaryotic cell[33]. The task is not easy[34]. But let me dwell on the terminology.

We have already reviewed some arguments that suggest that the problem of the position of Man in the cosmos depends critically on the evolution of microorganisms up to the level in which the emergence of nucleated cells ("eukaryogenesis") occurred. This forces upon us the position of the eukaryotic cell in the cosmos, as the main focus of our attention. Such a radical break with the past has some implications in our understanding of the origin and destiny of Man. Nevertheless, none of these arguments lie outside the scope of the question raised in the Papal Message to the Pontifical Academy of Science[35]. The Message to the Pontifical Academy is real progress for better understanding between science and theology:

> *With man we find ourselves in the presence of an ontological difference, an ontological leap, one could say. However, does not the posing of such ontological discontinuity run counter to that physical continuity which seems to be the main thread of research into evolution in the field of physics and chemistry? Consideration of the method used in the various branches of knowledge makes it possible to reconcile two points of view that would seem irreconcilable. The sciences of observation describe and measure the multiple manifestations of life with increasing precision and correlate them with the time line. The moment of transition to the spiritual cannot be the object of this kind of observation, which nevertheless can discover at the experimental level a series of very valuable signs indicating what is specific to the human being.*

In spite of this important step, the acceptance of evolution has not led to a consensus amongst scientists, either on its mechanism, or on its implications. Nevertheless, in spite of this shortcoming, in this book we base our arguments on Darwin's theory of evolution.

2.11 Questions in philosophy, science and theology

Constraints imposed by philosophy and theology on our view of life mostly favors a special place of man in the universe. In the case of philosophy there is a continued quest for the impact of technological progress on the future of mankind.

A relevant question concerns the changes in our theological outlook that may follow the incorporation of knowledge of the place of earthly biota in the cosmos. We should distinguish two levels at which life in the universe would influence the humanities.

On the one hand, we could have a universe full of microbial life, but there is a second possibility that the universe is full of intelligent life[36]. The discovery of widespread distribution of microorganisms in extrasolar planets would be of fundamental importance for appreciating whether the hypothesis of universality of biology is reasonable.

Philosophers and theologians have devoted much effort to the implications for our religious beliefs of intelligent life elsewhere in the universe. Thinking on this subject extends from Plato's *Timaeus,* a classic book, to a large body of literature published throughout last century. However, in spite of all the progress in astrobiology and other space sciences, there seems to be a prevalent feeling that widespread distribution of intelligent life is unlikely[37]. This view is not general, as we shall discuss in later chapters.

The present technological capacity to develop more powerful telescopes, and more ambitious space missions, as well as searching for signatures of life in other stars of our galaxy through different windows of the electromagnetic spectrum leaves open the possibility to escape from the view that the universe is only pregnant with microbial life except for the single case of life on Earth, a situation that would be reminiscent of a pre-Copernican anthropocentric view of the Earth.

Hence, in the presence of these uncertainties it is fair to ask:

Would there be problems in the traditional monotheistic view of Deity as being confined with the affairs of man?

It could be argued that the Deity is omnipotent and can be concerned with the affairs of all the intelligent species of the universe. If SETI (cf., Sec. 2.1), or a more restricted search in our own solar system, were to confront us with parallel evolution in other worlds, some intriguing questions would be raised. In this context, Steven Dick has coined the word "astrotheology"[36]. At any rate, the question of the impact of extraterrestrial life on our culture does define a substantial series of questions to merit a well-defined academic approach. Some of the deeper questions that humans have raised in the past are not always answerable within the boundaries of science. Instead, philosophers and theologians have approached such questions within their own domains of competence.

One example is provided by the question of purpose in evolution. Indeed, the concept of purpose in a general sense may be understood as something that one sets before oneself as an object to be attained, an aim to be kept in a plan. In attempting to give an answer to the question of "purpose in nature", we should discuss the main components of human knowledge in an integrated way, so as to ask the right questions in the right field of knowledge. This approach will induce us to provide appropriate answers that are reasonable within philosophy, science, or theology.

We wish to answer some questions in philosophy and theology that are pertinent to the main subject of this chapter. In order not to go beyond the natural boundaries of either science, or theology, we shall discuss contemporary attempts to encompass Darwinian evolution into natural theology. For the implementation of this objective, we need to discuss philosophical and theological issues.

2.12 The anthropic approach and the emergence of life

Teleology may be considered as either a doctrine according to which everything in the cosmos has been designed with Man in mind. Alternatively it can be interpreted as a theory of purposiveness in the cosmos. In other words, this theory maintains that phenomena should be explained in terms of its purpose, rather than by initial causes. This concept is thus intimately related to the Anthropic Principle, either in physics or biochemistry. We shall illustrate it below in the case of physics, where it first appeared.

Teleology is also related to various interpretations of fine-tuning in phenomena, such as the nuclear reaction of helium and beryllium atoms in the production of carbon, as originally pointed out by Sir Fred Hoyle[38]. His studies concerned nuclear phenomena related to the synthesis of some of the biogenic chemical elements inside stars such as our Sun. This phenomenon is especially remarkable for the elements carbon and oxygen. It was clear from his studies that if carbon is to be produced in the interior of stars, a series of coincidences had to occur with the physical entities that take part in the stability of the star. The need for carbon to occur in the interior of stars does not require the details of nuclear physics. It is sufficient to remark that gravitation tends to collapse the star onto itself. For this to be prevented four hydrogen atoms are fused into one atom of helium-4. But eventually, as the supply of hydrogen runs out and gravitational contraction increases, the density of the core of the star, its temperature is raised. Thereby its temperature increases to a point that the fusion of three helium atoms into an atom of carbon-12 takes place. But this occurs via an intermediate fusion of two helium atoms into one atom of beryllium-8. Subsequently the atom of beryllium-8 reacts with a third atom of helium-4 to lead to the final synthesis of the carbon-12 atom.

Of the several numerical coincidences that led Hoyle to point out the anthropic approach, we will only mention one. The lifetime of beryllium atom is a thousand times longer than the time that is required for the two atoms of helium to fuse into one atom of beryllium-8, which can therefore coexist with another helium atom for the reaction of beryllium and helium to yield the carbon-12 atom. At this stage we could say that the lifetime of beryllium-8 is "fine-tuned" for the conditions for carbon-based life to exist in the universe. What was even more striking in the thinking of Hoyle was to anticipate (before experimental confirmation) the type of reaction that had to take place for the initial three helium-4 atoms to produce carbon-12. (It is this type of thinking that was to be referred to as "anthropic".) Nuclear physicists call this specific phenomenon a resonant reaction. It simply refers to the existence of a series of narrow levels of energy (called resonant energy), at which the probability that the reaction will take place is greatly favored. It was discovered that there was a resonant energy for carbon-12 just above the energy of beryllium-8 plus helium-4, the missing energy for the reaction to take place was to be provided by thermal energy form the stellar interior[39].

A far-reaching implication of the possibility of interpreting the evolutionary aspects of Darwinism within theology is that the evolutionary process begins at the molecular level of biochemistry. In fact, such "chemical" evolution is a time-honored discipline that has been studied extensively in the past, particularly during the last decade of last century[40,41]. We have attempted to show that fine-tuning in biochemistry is a well-defined problem[42]. Its evolutionary aspects should in principle be able to be integrated into a framework of natural theology, for instance, in approaches such as kenotic process theology.

What is characteristic of *process* philosophy is not simply recognition of natural process as the building blocks of nature, but rather "process" is considered an essential aspect of nature. Process philosophy holds that what exists in nature is not just originated and sustained by processes, but in fact they characterize it. Alfred Whitehead (1861-1947) was an English mathematician, logician and philosopher. While he was working at Harvard from 1924 onwards, he developed a comprehensive theory of metaphysics, which that gave rise to process philosophy. Whitehead found necessary a God who is the giver of newness, a loving leader, intimately guiding, but never dominating, every experience[43]. Process philosophy, or "process thought", has been intended to provide a common metaphysical basis for discussions of science and religion. Some criticisms have been raised in the past[44]: if physics is to be appropriate for process thought, this school of philosophy has to face an ongoing debate; continuity seems to be intrinsic to quantum mechanics (for example, the Schrödinger equation is a differential equation). For Polkinghorne the mathematics of process thought ought to be that of difference equations, instead of differential equations (with their implied underlying continuity).

On the other hand, we should keep in mind deeper questions that are currently at a debatable stage: quantum gravity aims at a coherent theory of space-time. Space-time is a dynamical entity, and as such it would have quantum properties[45]. Both current and eventual developments in theoretical physics have to investigate the concept of *discrete* excitations of space itself. Thus process thought, as a philosophical system, cannot be ruled out at present by an unfinished debate[42].

2.13 Is there evidence of purpose in evolution?

We also discuss how the combined approach of philosophy, science, and ultimately natural theology can help us to approach the intelligibility of the universe in a rational way. Care is needed though to address the right questions within their corresponding cultural domains. First of all we consider the origin and evolution of intelligent behavior in the cosmos, as it already has taken place the in the case of life on Earth. We have already pointed out that there are ways of incorporating evolution of life on Earth (Darwinism) in terms of natural theology (i.e., process theology).

Our subsequent arguments in this chapter, within the boundaries of science, should be useful for their interpretation in terms of theological issues. In order to decide what is the position of our tree of life in the universe, we must appeal to science. First of all, we should put aside some philosophical objections that have been deeply rooted in the literature. We will come back to this topic in later chapters, when we consider constraints on chance and evolutionary convergence. Putting together these two aspects of evolution with its intrinsic randomness suggests the following possibility: There are trends within evolutionary history that might reflect the existence of general principles in the evolution of increasingly larger and more complex forms in the Earth biota, including the brain[46].

In previous chapters we have already mentioned Monod's book *Chance and Necessity*. The author overemphasized the role of "pure chance" in evolution. He excluded the role that evolutionary convergence may have had in the evolution of life on Earth. On this basis Monod concluded that such trends in biological evolution must be rejected[47]. This question is not merely philosophical, although its philosophical implications are important. The question of evolutionary trends is relevant to the subject of astrobiology and in particular to bioastronomy. There has been an enormous technological revolution in the capability of scanning the celestial sphere for traces of ongoing communication amongst creatures that are the product of evolution of intelligent behavior elsewhere. In concluding this chapter we underline the fact that chance at the molecular level in terms of mutations in the genome, does not exclude organisms from exhibiting trends at a higher level of organization.

Another aspect of the question of the existence of trends in evolution is also relevant to natural theology, and has been discussed extensively[48]. This seems to be an appropriate place to leave the subject of deeper implications of the search for extraterrestrial life at this particular crossroad of bioastronomy, philosophy and theology. However, we would like to conclude this chapter by underlining the importance of the trends in evolution for both science and theology.

Firstly we consider science: the rationale for the SETI project is to assume that trends in evolution that have been observed on Earth, may serve as a basis for understanding the eventual "contact" between different forms of intelligent creatures that do not belong to the same tree of life[49].

Secondly, the trends that have been observed in evolution on Earth may serve for the intrinsic and necessary problem in theology, namely, rationalizing the concept of Divine action, without the fear of not being able to establish a reasonable constructive dialogue with science. The realization that randomness in evolution does not rule out the possible existence of trends in evolution opens the door for real progress in the integrated approach to all forms of culture. Such an approach will not fall in the trap that dates back to the publication of Darwin's seminal work, when possibly, because of the difficulty of communication between science and religion, a confrontation between faith and reason emerged. Unfortunately, such a confrontation is still with us.

Chapter 3

From a First to a Second Genesis

In Chapter 2 we saw how the search for life in the universe fits into the development of Western civilization. Any account of the philosophical issues of the theory of evolution will be relevant to astrobiology. For this reason we highlight reductionism, which refers to any doctrine that claims to reduce the apparently more complex phenomena to the less so[1]. The successes of reductionism in science are well known to scientists and the educated laymen.

The reader is advised to refer especially to the following entries in the Glossary: Augustine of Hippo, Design Argument, DNA, Henderson, Miller, Natural Theology, Optical activity Pasteur, Reductionism, Revelation and RNA.

3.1 The fitness of the environment

We continue deepening our appreciation of how Western civilization has led to the present intimate connections between the evolution of the universe and the phenomenon of life. The fitness of the cosmos for the origin and evolution of life is discussed in Lawrence J. Henderson's *"The Fitness of the Environment"*. This influential book was published over 90 years ago. Hend' main interests ranged widely: he was a physiologist, chemist, biologist, philosopher and sociologist. He discussed the question of teleology in biochemistry to give some rationale to the question of fitness of the environment for the evolution of life. For many chemical compounds he discussed the difficulties that the evolution of life would have encountered had these compounds not been freely available in the environment. Water was one example. Its search, even today, is a main objective of the exploration of the Solar System. Henderson concludes at the end of his book that:

> *"The properties of matter and the course of cosmic evolution are now seen to be intimately related to the structure of the living being and to its activities; they become, therefore, far more important in biology than has been previously suspected. For the whole evolutionary process, both cosmic and organic, is one, and the biologist may now rightly regard the universe in its very essence as biocentric."*

25

Today we should search the roots of Henderson's "biocentrism" at the molecular level. In fact, fine-tuning in biochemistry is represented by the strength of the chemical bonds that gives rise to the "dictionary" that contains the correspondence between amino acids and the monomers of RNA. This dictionary is called the genetic code. Both transcription and translation of the messages coded in RNA and DNA would not be possible if the strength of the bonds had different values. Hence life, as we understand it today, would not have arisen.

We shall argue in favor of fitness of the cosmos for the origin and evolution of life without touching on the question of teleology. In this sense, we approach the subject without restricting ourselves exclusively to biological evolution in the universe, but rather we also include the evolution of the structure of the cosmos itself.

We shall touch upon the evolution of solar systems, evolution of interstellar matter and, finally, various aspects of evolution of the cosmos—all in relation with the emergence of life. However, we hasten to point out that arguments based on science can, nevertheless, be a source of inspiration for reconsidering the bases of natural theology. In Chapter 7 we shall argue that fitness of the universe for the origin and evolution of life can be best understood, not only through convergence in biochemistry, but also through a range of convergences in the space sciences.

3.2 Life appeared spontaneously in the remote past

The research of Louis Pasteur included some of the most remarkable insights in medicine, as well as in what we now call optical activity. He had brought upon the scientific community the urgency of rationalizing the question of the origin of life. It became evident, due to Pasteur's own work, that the concept of spontaneous generation was untenable. In 1860 Pasteur wrote to a friend on the "impenetrable mystery of life and death"[3]:

> *There is so much passion and so much obscurity on both sides that it will require nothing less than the cogency of an arithmetical demonstration to convince my adversaries of my conclusions.*

When Pasteur was writing these lines, Charles Darwin had just published "*The Origin of Species*". It was only with the publication of this fundamental work that the basic questions on the nature of life would be seen in their proper perspective. In fact, the fundamental question of "how life might have been breathed into matter", had to wait for two events that were to be well separated in time.

First of all, Darwin had to have his work widely read and discussed by the scientific community. The publication of his theory of evolution had taken place the previous year, although it had been maturing ever since his trip on *The Beagle*. Darwin's publication was motivated by Alfred Russell Wallace (1823-1913), whose work was the fruit of travels as a naturalist on the Amazon and in the Indo-Australian archipelago. Wallace independently had developed a version of evolution by natural selection.

Secondly, the first steps in the understanding of the spontaneous appearance of life from basic chemical compounds had to wait for another hundred years for the emergence of pioneering works of chemical evolution (cf., Sec. 8.4). Indeed, it was not until the advent of chemical evolution itself in the middle of last century, when scientists suggested the possibility of life spontaneously appearing on Earth somewhere in the remote past. On the basis of the work of Darwin, Wallace, Oparin and Miller philosophers were to be in a position to discuss one of the deepest questions that science has raised up to the present time, namely,

What is life and how did it first appear on Earth?

3.3 Philosophical issues related to life in the universe

The success of science during the 17th century can probably be best illustrated with Isaac Newton's theory of gravitation; with the introduction of quantum mechanics by Max Planck, and finally with the unified electroweak theory of Sheldon Glashow, Abdus Salam and Steven Weinberg. On the other hand, there is a second meaning for the word reductionism[4]: it is the belief that human behavior can be reduced to or interpreted in terms of the lower animals, and that ultimately it can be reduced to the physical laws controlling the behavior of inanimate matter. For convenience this form of reductionism will be referred in this text as its strong version. For example, sociobiologists such as Edward Wilson have followed the tradition of the giants of physics, who restricted themselves to the weak form of reductionism). These groups of life scientists have extrapolated reductionism into its strong form. Indeed, the basis of human social behavior has been studied, in order to determine the relation between genetic constraints and their cultural expression.

Some opponents have even referred to this approach as "genocentrism". Further applications of strong reductionism have led to controversy, as clearly illustrated by Robert Russell in his review on

"Life in the universe: Philosophical and Theological Issues"[4].

In particular, Russell discusses the arguments that have been put forward in favor of interpreting the capacity and content of human morality as products of evolution.

3.4 The question of design in biology

The question of design in biology has a long history going back to ancient Greece. However, in modern times we may begin with the work of William Paley (1743-1805). He was an archdeacon, and Doctor of Divinity at Cambridge University. His writings were highly respected in the Anglican order. His *Horae Paulinae* (1790) was written specifically to prove the historicity of the New Testament. Another famous book was *View of the Evidences of Christianity* (1794), a text that was standard reading amongst undergraduates during Charles Darwin early university education. However, his best-remembered book is *Natural Theology,* which played an important role in the early stages of the establishment of Darwin's arguments. Paley presented some observations from nature that was meant to prove not only the existence of a grand design, but more importantly, in his "Natural Theology" Paley attempted to prove the existence of an intelligent designer. The famous quotation is at the beginning of his book:

> *Suppose I found a watch upon the ground, and it should be enquired how the watch happened to be in that place...When we come to inspect the watch, we perceive that its several parts are framed and put together for a purpose...the inference, we think, is inevitable, that the watch must have had a maker...*

This argument can be traced back to classical times, but Paley's defense of it in modern times was influential in the 19th century dialogue between science, philosophy and theology. One of the fundamental steps in the ascent on man towards an understanding of his position in the universe (a key to understanding the present state of astrobiology) has been the realization that natural selection is indeed a creative process that can account for the appearance of genuine novelty, independent of a single act of creation, but more as a gradual accumulation of small successes in the evolution of living organisms.

This is a point that has been defended by the founders of Darwinism[5]. Francisco Ayala refers to an analogy with artistic creation. The creative power of natural selection arises, according to Jacques Monod[6] to an interaction between chance and necessity, a phrase that became familiar thanks to his very popular book "*Chance and Necessity*". Consider a painter who mixes and distributes pigments over a canvas. A random mixture of pigments could not create Leonardo's Mona

Lisa, or at least the probability is infinitesimally small. This underlines the fact that natural selection is like the painter. It is not a random process. The complicated anatomy of the human eye, for instance, is the result of natural selection.

3.5 Towards a general interest in the origin of life

Theologians have been deterred to get involved in questions related to our origins based on reason, rather than revelation. Some reservations can be traced back a long time to the fundamental question of how to read the Holy Books of the three monotheistic religions of the world, namely the Old Testament of the Holy Bible that is shared by Judaism, Christianity and Islam.

In the Book of Genesis there are questions raised that are of interest to theologians, philosophers, scientists and artists. To extend the fortunate phrase of Lord Snow[7], we may refer to these four groups as the "four cultures". In this terminology, it is hardly surprising that the fourth culture (art) should have shown interest in Genesis. The very rich iconography of Christianity and Judaism was a permanent source of funding for artists throughout the rise of Western civilization. (Our book cover is a good illustration of his point.) The reason why the third culture (science) has been interested in the evolution of life is self-evident. Science took one of its most transcendental steps towards understanding the complexity of the biosphere when Darwin and Wallace formulated natural selection as a mechanism for evolution. Progress since the publication of *The Origin of Species* has been considerable. The second culture, philosophy, has been intricately connected with the development of Darwinism. Philosophers such as Karl Popper and John Dewey have meditated implications of the Theory of Evolution.

We wish to dwell at a certain length on the interest that the first culture, theology, has had on the question of the origin and evolution of life on Earth and the possibility of there being life elsewhere in the universe. Traditionally, there has been a certain caution of theologians with respect to the questions that we have discussed at some length in this work. Saint Augustine touched on this question in *The City of God* with respect to possible conflicts that may arise from a literal reading of the Bible[8]. In our own time a clear position was expressed at the Pontifical Academy of Sciences, which had met to discuss the origin and evolution of life. In this message, the subject of this book was defined as John Paul II[9]:

> *"a basic theme that greatly interests the Church, as revelation contains, for its part, teachings concerning the nature and origins of man".*

Raising the Augustinian conundrum of whether scientifically reached conclusions, and those contained in revelation on the origin of life, seem in contradiction with each other:

> *"In what direction should we seek their solution? We know in effect that truth cannot contradict truth".*

This position has opened the way to a fruitful dialogue between scientists and theologians, two cultures, which are not too distant from each other[10]. New understanding is arising between cultures that instead were far apart.

3.6 Cultural implications of life in the cosmos

Recalling the two main subtopics of astrobiology that we have been discussing in previous pages, namely, distribution and destiny of life in the universe, what we have argued is that if future missions for the exploration of the Solar System were successful in the identification of a second Genesis, the science of the distribution of life in the universe would lie on solid scientific bases. This concludes the first of the two topics I wished to comment upon. Given such bases for the distribution of life in the universe, it does not seem premature to include in our discussions other sectors of our society. Keeping this in mind, we move on to the second topic of the present work, the destiny of life in the universe. Indeed, in the future, awareness of other life forms would have a significant impact in our culture, not just in our science. The influence of the new knowledge will also be felt while discussing questions on our origins, not only against a background of our particular evolutionary line, which has been followed up by life on Earth. Such questions ought to be discussed Rather than having just one example of a tree of life, we would be in the presence of a whole "forest of life", that is many evolutionary lines.

In fact, the intercultural dialogue is urgent for various reasons, which are strongly suggested by astrobiology:

• A human-level type of intelligent behavior may be widespread in the cosmos.
• Within the context of astrobiology it is clear that our human descent does go back all the way to microorganisms and,
• Ultimately, our origins go back to star dust.

Yet, these three items lie outside our cultural patrimony. But it is not an easy task to integrate our knowledge. In this work, my main overall thesis has been that such an integration is not only possible, but it is also timely, and urgent.

Chapter 4

Implications of a Second Genesis

In Chapter 3 we discussed some implications of a second Genesis. Now we enquire whether it is compatible with faith and reason. An answer to this question requires special care: the compatibility of a second Genesis with science is the subject matter of astrobiology, while the compatibility of a second Genesis with religious belief lies within the domain of natural theology. Monotheism does not present us with any particular stumbling stone for incorporating the emergence of life beyond the Earth within religious traditions. This issue has implications on theoretical and moral philosophy. Our discussion will remain strictly within the boundary of science. We ask whether the evolution of intelligent behavior is inevitable, given what is known in the new science of astrobiology.

The reader is advised to refer especially to the following entries in the Glossary: Anthropic Principles, Biogeocentrism, Biomarker, Chardin, Christian, Convergent evolution, Crick, Divine action, Enlightenment, Foundationalism, Haught, Maritain, Platinga and all entries on "Process".

4.1 Convergence of the evolutionary process

The concept of the fitness of the cosmos for the origin, evolution and distribution of life is relevant for our discussion of the compatibility of a second Genesis with science and religion. As mentioned in Chapter 3 the American scientist and philosopher Lawrence Joseph Henderson introduced this concept in science at the beginning of the 20th century with a more limited scope[1]. This issue has implications on theoretical and moral philosophy, but we are particularly concerned with implications that astrobiology may present to natural theology. The topic of the relationships between science and religion has been discussed in depth particularly since the Enlightenment[1].

Our discussion will remain within the natural boundary of science. We are concerned with the question of whether the evolution of intelligent behavior is inevitable, given what we know from the new science of astrobiology. In his book *Religion and Science* Bertrand Russell isolated the main problem at the basis of the question whether our universe is fit for the emergence and evolution of life. At a time when Alexander Ivanovich Oparin was struggling to understand the origin of life on Earth, Russell wrote[2]:

"The three centuries that have elapsed since the Giordano Bruno suffered martyrdom for believing in the plurality of worlds have changed our conception of the universe almost beyond description, but they have not brought us appreciably nearer to understanding the relation of life to the universe."

Astrobiology is a broad discipline that can guide us into an understanding of life in the universe. In fact, it is the correct tool for the search of the answers that science can provide regarding the question of the fitness of the universe for the emergence and evolution of life. In this context we can attempt to formulate questions that are of interest beyond science:

Does complex chemistry offer evidence of purpose?

The intimately related questions of the Anthropic Principle and fine-tuning in living systems[2] are questions that would be simpler to understand with more than a single Genesis. On the other hand, our religious traditions go back to Jewish theology: there is a sole omnipotent God who created heaven and earth, and subsequently life on earth. This view of our origins has traditionally been referred to as (a "first") Genesis. But revelation through the scriptures never raises directly the question of the plurality of inhabited worlds. There is no incompatibility between religious tradition and the possibility that we may not be alone in the universe. What is exciting about the emergence of the new science of astrobiology is that we can explore the possibility that the evolution of intelligent behavior may be inevitable in an evolving cosmos. In Chapter 1 we have referred to the possibility of life originating elsewhere in the universe as a "second Genesis"[3]. An aspect of these reflections should be highlighted from the beginning: Our lives are short and we crave for an answer to the question: *Are we alone in the universe?* No intelligent signals have been identified after almost half a century of searching for life in the universe. This investigation has been carried out through windows of the electromagnetic spectrum. Nevertheless it should be emphasized that considerable technological progress has been achieved in the techniques being used since the SETI project began in the middle of last century. But technology has not been the only field of the space sciences that has progressed in recent years. The exploration of the solar system has also been remarkable with a fleet of missions that are capable of detecting microscopic life.

The search for extraterrestrial life was attempted for the first time on the surface of Mars a quarter of a century ago. The Viking missions were capable of detecting microbial life. Sadly, the results were not convincing to most scientists. The search still continues today with Mars being the present target of several space missions from NASA, ESA and Japan. Yet given the harsh conditions for the survival of extremophilic microorganisms on the Red Planet, the best digging equipment with present technology is still unable to probe as far as the more likely sites, deep underground, where we expect abundant liquid water to be present.

4.2 Is biology sufficient to discuss a second Genesis?

In order to consider the question whether a second Genesis is possible, we should first decide if the science of biology would be the right tool to apply in order to find out whether there is life anywhere in the universe. Indeed, some issues have been discussed in the past regarding the universal nature of biology in general, and biochemistry in particular: firstly, life may be a cosmic imperative[4]. A somewhat different approach is due to Francis Crick[5]: In the "directed panspermia" hypothesis life can emerge in some solar systems by directly sending microbial organisms to barren planetary or satellite environments.

Secondly, multicellular life may be a rare phenomenon in the cosmos, although the existence of microbial life may still be widespread[6]. Finally, a third related issue is the possibility that evolution of intelligent behavior may be just a question of time (and preservation of steady planetary conditions), and hence ubiquitous in the universe. Darwin's theory of evolution is assumed to account adequately for the phenomena that we associate with life anywhere in the universe[7].

We argue in favor of the inevitability of the origin and evolution of life. We assume that Darwinian evolution is a universal process and that the role of contingency has to be seen in the restricted context of parallelism and evolutionary convergence[8]. Convergence is not restricted to biology, but it may also be extended to other realms of science. The question[9]:

"What would be conserved if the tape of evolution were played twice?"

has been raised repeatedly in the past[9]. Since all forms of life known to us are terrestrial organisms, it is relevant to the question of whether the science of biology is of universal validity[10, 11]. Independent of historical contingency, natural selection is powerful enough for organisms living in similar environments to be shaped to similar ends. For this reason, it is important to document the phenomenon of evolutionary convergence at all levels, in the ascent from stardust to brain evolution. In particular, documenting evolutionary convergence at the molecular level is the first step in this direction. Our examples militate in favor of assuming that, to a certain extent and in certain conditions, natural selection may be stronger than chance[12,13]. Biochemistry, a science supporting chemical evolution, is assumed to be a universal science, not just a science applicable to the Earth biota.

4.3 Is fine-tuning related to a second Genesis?

We have assumed that natural selection seems to be powerful enough to shape terrestrial organisms to similar ends, independent of historical contingency. Evolutionary convergence can be viewed as a "re-run of the tape of evolution", with final results that are broadly predictable (we shall return to this analogy later on). Hence, if life arises again elsewhere in the cosmos, we would expect some degree of convergence with the evolution of terrestrial human-level of intelligence. One scientific approach to test the hypothesis of a second Genesis outside the Solar System is to search for extraterrestrial intelligence (referred to in Chapter 3 as the SETI project[14,15]).

One of the objectives of SETI is to test whether trends in evolution that have been observed on Earth may serve as a basis for understanding the eventual "contact" between different forms of civilizations that do not belong to the same tree of life. Natural theology, on the other hand, is the body of knowledge about religion that can be obtained by human reason alone without appealing to revelation. Trends towards larger brains that have been observed in evolution on Earth, may serve for rationalizing the concept of Divine action in a constructive dialogue with science. The realization that randomness in evolution does not rule out the existence of evolutionary convergence encourages an integrated approach for science and religion. Such an approach will clearly avoid a confrontation.

We should decide whether the evolution of intelligent behavior has followed convergent evolutionary pathways. In this context, we can begin testing the lowest stages of the evolutionary pathway within the Solar System. We are in a position to search directly evolutionary biomarkers in the Jovian moon Europa. We have considered a set of evolutionary biomarkers if extant microorganisms are to be encountered on Europa[16].

Why do we focus our attention on this satellite of Jupiter? There are several reasons, but the main one is related to the results retrieved from Galileo, the most important mission for the exploration of the Solar System that took place last century. The spacecraft arrived in the Jovian system in 1995 and completed its campaign in the month of September 2003. This mission has exposed an environment that can, in principle, support life[17]. However, the most interesting cases will have to wait for the future. Eventually, we should be able to determine specific locations where the icy surface is thin enough for a submersible penetration or, if the icy crust proves to be too thick, for testing directly on the surface for the presence of microorganisms. We may conclude that within the realm of scientific research in the foreseeable future we can address the question of "fine-tuning": evolution of the cosmos, and especially biological evolution right from the biochemical level, may be "fine-tuned" for the inevitable emergence of intelligent behavior in the cosmos, provided there is preservation of steady planetary conditions over geologic time.

4.4 Reflections by theologians

One of the main ingredients of the study of life in the universe is convergence of the evolutionary process in biology. The evolution of life in the universe, either microscopic, or even life at a human-level of intelligence, presents no insurmountable difficulties to natural theology. Witness to this fact is the statement made by Pope John Paul II in the presence of the Pontifical Academy less than a decade ago[18].

Science and religion are both concerned with the common understanding of the destiny of life in the universe. Since these two intrinsically different cultural activities largely address the same questions, they should, at some point, establish a dialogue, since the search for truth from different points of view should inevitably lead to a common objective.

What may present more of a conflict between science and religion is not the confrontation of science with a second Genesis. Instead, a real difficulty would be the evolution of all the attributes of man, including those that are of prime importance for theology—the spirit of man. The question has been formulated a little more precisely[19]:

> *Is a creationist theory required to explain the origins of*
> *the spiritual dimension of the human being?*

While we are still not in a position to answer this question, in Chapter 12 we attempt to explain that contact with extraterrestrial life cannot be excluded in the foreseeable future. Knowledge, or merely awareness, of a second Genesis would provide us with a solid point of reference on which to base discussions on the implications of all the attributes of beings that have evolved to a human-level of intelligent behavior. In such discussions the participants should be scientists and natural theologians.

Yet we wish to remain within the constraints that the seminal contribution of Galileo Galilei has imposed on us. Science must remain as an experimental academic activity. Hence, the question of man's spirit and soul should, in principle, not even enter the biological discourse. However, man's spirit and soul are concepts that are relevant to moral philosophy (ethics). It may be argued that ethics and other branches of human knowledge should be integrated. The integration of biology and ethics seems particularly relevant to the dialogue of science and religion[20,21].

Efforts towards such integration will undoubtedly help a coherent discussion of the evolution of the attributes of man. Such inquiries should include specific attributes of man that are most relevant from the point of view of theology. The subject of the philosophical and theological implications of a second Genesis is an open problem[22]. The large number of papers, books and encyclopedia articles devoted to this subject demonstrates this assertion. The authors of these papers have been philosophers, scientists and theologians[23-28]. It is remarkable that a substantial number of theological discussions appeared in the middle of the 20th century, although the subject itself of the possible role that is played by the Creator in the distribution of life in the universe began at least over two millennia ago. In fact, more than 2300 years ago humanists have speculated about the possibility that the maker of the universe (Plato, 360 BC)[29]:

"distributed souls equal in number to the stars, and assigned each soul to a star."

Several theologians preceded the current wave of enthusiasm on the topic of the nature of extraterrestrial beings that may have reached a human-level of intelligence. The authors of these papers that were published since the 1960s were: T. J. Zubek, John P. Kleinz, James Harford, Daniel C. Raible, George Dugan, A. Carr, John J. Lynch, L. C. McHugh, Angelo Perego, Joseph A. Breig and J. Edgar Bruns (for a more detailed bibliographic information, we refer the reader to the paper of Douglas Vakoch[30]; for references to the work of previous centuries we refer the reader to the review of the astronomer and historian of science Francesco Bertola[31].

A question that has been repeatedly raised in the past is the following:

Should Christians expect that a single Incarnation of Christ in
Jesus is sufficient for the redemption of all life in the universe?

This question has been discussed previously: Davis suggested that it might be unnecessary to postulate additional Incarnations[23]. Man's reconciliation with God through the sacrificial death of Christ is assumed for understanding the redemption of any beings that may have evolved to a human-level of intelligence elsewhere in the universe. Other authors agree with such a single, universal Incarnation, for instance, Ted Peters[32] and the distinguished English cosmologist and philosopher Edward Arthur Milne (1896-1950)[33].

Some of the barriers to the dialogue between science and religion can be traced back, to the fundamental question of how to read the Holy Books of the three monotheistic religions. In the 5th century, Augustine had touched on this question in his book in *"The City of God"*[34]. This question is discussed in more detail in Sec. 12.1.

If science were to provide irrefutable evidence of the emergence of life elsewhere, theologians could read the texts of the New Testament in a broader context[28] (Jn 10:16):

> *And I have other sheep that are not of the fold; I must bring them also, and they will heed my voice. So there shall be one flock, one shepherd.*

On the other hand, an alternative view in natural theology, envisages multiple Incarnations. This approach has also received much attention. Christian theologians in this group are Paul Tillich[35], Eric Mascall[36] and Ernan McMullin[37].

Finally, Pierre Teilhard de Chardin (1881-1955), the French philosopher and paleontologist, developed an interdisciplinary approach[38] to the question of redemption that has implications on the evolution of life in the universe. Medawar[39] has criticized it. However, Teilhard's work should be seen not as an alternative to scientific thought, but as a personal effort to achieve a synthesis of his mainly theological thinking. Teilhard turned away from an exclusively scientific, or even an exclusively philosophical work, when he accepted to convert his work into a contribution to natural theology. (In the past some of the critics of *"The Phenomenon of Man"* have ignored the restricted scope of Teilhard's natural theology[40].)

Indeed, he suggested that when humanity and the world have reached their final state of evolution, a new convergence between them and God would be initiated by a future return of Christ to judge both the living and the dead (a "Second Coming" of Christ). Teilhard asserted that the work of Jesus of Nazareth was primarily to lead the material world to a truly cosmic redemption.

4.5 Naturalism in philosophy

Naturalism in philosophy is a concept used by G. E. Moore as an approach that intends to relate the scientific method to philosophy by affirming that all beings and events in the universe are natural. Consequently, all knowledge, including ethics, falls within the range of scientific investigation. The American philosopher John Dewey[20] was a strong supporter of this doctrine.

On the other hand, Jacques Maritain emphasized the notion of a "cosmic morality" that is independent of Darwinian evolution. His review and criticism of the views on naturalism of John Dewey is relevant if it is inserted in the context of the possible existence of life in the universe. Sadly, Maritain passed away in 1973. Since that time much progress has taken place in natural theology to sort out evolutionary biology's challenge to theology; in other words, the puzzle of God's relation to an evolutionary process that is characterized by the long history of variation, mutation and selection[41]. Some clarification of these issues has been formulated by the contemporary American theologian John Haught, who has discussed the compatibility of Darwinism and natural theology in the context of kenotic process theology[42] (cf., Sec. 12.7 for a more detailed discussion).

A few words are needed to introduce the concept of "process" in this context. Indeed, process philosophy holds that what exists in nature is not just originated and sustained by processes, but in fact they characterize it. The original discussion is due to the English mathematician and philosopher Alfred North Whitehead, a Harvard University professor who developed a comprehensive metaphysical theory.

These ideas were extended into process theology an approach that rejects Divine action in terms of causality, proposing that God acts persuasively in all events, but not necessarily determining their character. It was during his stay at Harvard that he worked on more general issues in philosophy, including the development of the comprehensive metaphysical system that has come to be known as process philosophy. He was the author of the influential books such as *"The Concept of Nature"* published in 1920 and *"Process and Reality"* published nine years later

On the other hand, kenosis means self-emptying and voluntary sacrifice on behalf of others[43]. A philosophy like John Dewey's draws our attention to the possibility that ethics may be regulated exclusively by the "positive sciences", namely the sciences that range from physics to biology and sociology. In Dewey's integrated view of science and philosophy, it is assumed that the above-mentioned scientific disciplines should be comprehensive enough to allow intelligent beings to make ethical choices according to scientific procedures.

Jacques Maritain was under the influence of the philosopher Henri Bergson. He insisted on the idea that moral philosophy (ethics) should take into account other branches of human knowledge. Maritain distinguishes the concepts of spirit (a theological concept) and nature (a scientific subject). In spite of this Maritain argues that there is room for the supernatural, as Christianity understands this word, for instance, in the interpretation of the Gospels, or more generally in the interpretation of the Judeo-Christian Bible and the Holy Koran[21]. A criticism that has been directed to the views of Maritain, is that his arguments are based on a form of foundationalism. This term means that knowledge could be started from basic beliefs (which in turn may support other beliefs, thus providing a "foundation" upon which all new knowledge could be inferred). Such basic beliefs are assumed to be self-evident; they need not be justified by more basic beliefs. There are two different ways beliefs are justified[44]: some beliefs are justified by being based on evidence, while other beliefs are justified even though they are not based on evidence. This two-fold division of justified beliefs forms the basis for foundationalism. Clearly, in Maritain's work and, more generally in Christian theology, some form of foundationalism may be expected to be inevitable.

4.6 Non-foundational premises of Christian belief

It is important to recall arguments in favor of the viability of Christian belief on non-foundational premises due to the work of the American contemporary theologian Alvin Plantinga. The focus of the defense of his philosophical system, known as Reformed Epistemology[44-47], is to concentrate on the question: Does basic theistic belief count as knowledge if it is true? Here "theistic belief" means any belief that directly entails the existence of the God of monotheistic religions. The backbone of a philosophical system that attempts to equate theistic belief with knowledge is the understanding of a postulated property that converts a true belief into knowledge. This property has been labeled as "warrant", thus the philosophical system of Reform epistemology is strongly steeped with the development of what warrant really is in the specific case of this philosophical doctrine. The English language usage of this word is "something serving as a ground for a belief", for instance, as used in the phrase "this development gives warrant in saying that it is new".

Thus Warranted Christian Belief is a philosophical approach attempting to replace foundationalism. Briefly, "evidentialism" is the doctrine that claims that theistic belief must be based on *evidence* in order to be justified. Plantinga argues that when we define evidentialism as the view that theistic belief is justified only if it is based on other beliefs, it turns out that evidentialism is indeed the denial of Reformed Epistemology. Plantinga discusses the relationship between evidentialism and foundationalism and argues against them. This approach opens another possible discussion of other intelligent beings on different lines from Maritain.

Although some of the views of Maritain can be criticized, especially from the point of view of the Reformed epistemology, the attributes that universal biological evolution may grant to other intelligent beings are still relevant today for the scientific search of extraterrestrial intelligence (SETI, cf., Sec. 2.1). How far can we trust that evolution will give convergent results on attributes that we already know to have evolved in human beings?

We do not wish to address the question of whether the position highlighted by Dewey is tenable, or whether the opposing view of Maritain is valid, but the above arguments, and some of their criticisms, are presented merely as illustrations of the relevance of a second Genesis to human culture, not only to the new science of astrobiology, but also to both moral philosophy, as well as natural theology.

To sum up, up to the present time the intelligibility of the universe has been a topic restricted to natural theology. The arguments presented in this chapter argue in favor of bringing this fundamental topic within the frontier of science. Consequently, future exploration of the Solar System, and beyond, in the search for other lines of biological evolution, should be considered a priority in science, as well as in ethics and natural theology.

Chapter 5

The Destiny of the Universe

In Chapter 4 we discussed implications of the presence of life elsewhere in the universe. Using some concepts of physics we now refer to the question of the Anthropic Principle in an attempt to render intelligible one aspect of the emergence of life. Although different solar systems are relatively distant from each other, nevertheless, the mechanisms that lead to life may be of a universal nature.

The reader is advised to refer especially to the following entries in the Glossary: Almagest, Anthropic, Aquinas, Aristarchus, Aristotle, Baryon, Bruno, Convergent evolution, Cosmogony, Cusa, Digges, Friedmann, Hubble, Natural selection, Penzias, Prime Mover, WMAP and Wilson.

5.1 The concept of convergence in the life sciences

In Sec. 2.12 we considered the origin of the Anthropic Principle. Such questions on "fine-tuning" in physics have led to a weaker and a stronger version of the Anthropic Principle, which is concerned with the existence of life, particularly intelligent life, in the cosmos. There is no difficulty in accepting what has come to be known as a "Weak" Anthropic Principle in physics:

Change the laws (and constants of nature) and the universe that would emerge most likely would not be compatible with life.

In biochemistry there is clearly an analogous statement:

Omit firstly the observed cosmic abundance of the biogenic elements that are favorable to life. Secondly, omit the environments (Earth-like planets, or Europa-like satellites) that favor evolution and adaptive radiation. The consequence of the omission of both of these factors is that most likely life would not arise.

However, we could in principle go beyond the frontiers of science in both cosmology as well as biology, by allowing some degree of teleology to be brought into the argument. We are referring to the formulation of the anthropic principles in the following "strong" terms:

- The laws of nature and the physical constants were established so that human beings would arise in the universe.
- The distribution of Earth-like environments and Europa-like satellites were laid out so that human beings would arise in the universe.

The general mechanisms of nature, according to the evidence that we can infer from the Earth biota that by now is at least 3 billion years old, imply that the evolution of intelligent behavior seems inevitable. What is not evident is the inevitability of the emergence of human beings. The intimately related questions of the Anthropic Principle and fine-tuning in living systems[1] are topics that would be simpler to understand with knowledge of more than a single instance of emergence of life on planet Earth. On the other hand, our religious traditions go back to Jewish theology: there is a sole omnipotent God who created Heaven and Earth, and subsequently life on Earth.

This view of our origins has traditionally been referred to as (a "first") Genesis. As mentioned in Sec. 1.3 with the emergence of astrobiology[2], we can explore the possibility of the occurrence of a "second" Genesis, namely, whether the evolution of intelligent behavior is inevitable in an evolving cosmos, given the present laws of cosmology (General Relativity) and the general mechanisms of biological evolution (natural selection and adaptive radiation). If we were to change these laws and mechanisms, the arguments supporting the inevitability of the evolution of intelligent behavior would not stand and thus the evolution of intelligent beings would not necessarily take place. This aspect of evolution is relevant to a Weak Anthropic Principle.

In the rest of this chapter we shall review the emergence from philosophy and theology of some relevant aspects of science that are needed for understanding what is involved in the discussion of the search of evidence for a second Genesis. The early Christians were faced with the necessity to respond to the insertion of the new ideas of Christianity in the context of classical Greek philosophy. At an early stage in the growth of Christianity within the Roman Empire, it was realized that in order to convert non-believers, the Christian message had to be adapted to current philosophical systems. At this stage the early clerics, particularly the monks contributed to the preservation of ancient culture by copying the old manuscripts of Greece and preserving them in their libraries. In the Byzantine Empire the ancient traditions continued.

The following thousand years that followed the collapse of the Roman Empire saw some growth in the disciplines that the Holy Koran encouraged. An example is the progress in astronomy. The "dark-ages" were instead an era of considerable progress.

5.2 Intelligibility of the cosmos according to Aristotle

To fully appreciate the concepts that are behind the discussion of a second Genesis we must return to the dawn of civilization in Ionia, an ancient region comprising the central sector of the western coast of Anatolia (now in Turkey). Ionia consisted of the northern territories that also included the city of Athens. Its geography in the Eastern Mediterranean placed it in a strategic position that allowed it to become a maritime power Economic prosperity, as in many other periods in history, led to progress in the cultural area. We are especially concerned with the birth and early evolution of philosophy. The most influential of all the cities of Ionia was Miletus (cf., Sec. 2.3). By the end of the 7th century the Miletus and other Ionian cities had achieved great prosperity through trading and colonization. Thales emerged during this period, as discussed in Sec. 2.3.

Two centuries later Aristarchus of Samos (310-230 BC) one of the last of the Milesian School that had founded studies of philosophy anticipated a fairly modern cosmology. Aristarchus already formulated a complete Copernican hypothesis, according to which the Earth and other planets revolve round the Sun (heliocentric theory); and in so doing, Aristarchus asserted, the Earth rotates on its axis once every 24 hours. In spite of this deep insight, the heliocentric theory did not prosper in antiquity, possibly because he lived at a time when more influential philosophers were also trying to make the universe comprehensible. A work usually attributed to Aristarchus that has survived to the present time, *On the Sizes and Distances of the Sun and Moon*, is based on a geocentric worldview.

Aristotle (384-322 BC) was a Greek philosopher and scientist. Aristotle, (together with Plato and Socrates), is considered amongst the most distinguished intellectual of Ancient Greece. Aristotle is one of the most important pioneers in Western philosophy. At an early age, he went to Plato's Academy in Athens. Remaining there until after Plato's death some 20 years later. Aristotle's views on physics influenced medieval science, until the birth of modern physics with Copernicus and his contemporaries. The term a "Renaissance man" applies accurately to Aristotle for his interests ranging from physics to chemistry, biology, zoology, botany, psychology, political theory, ethics, logic and metaphysics, history and literary theory.

His philosophy supported both early Christian, as well as Islamic scholastic developments. Aristotle was a contemporary of Aristarchus. Driven by the just-mentioned all-embracing interest in human culture, not surprisingly Aristotle attempted to make the universe intelligible by formulating a complete cosmology. Since observational astronomy was at a primitive stage, he succeeded in Western civilization's first attempt to demonstrate the intelligibility of the universe. Later on, with somewhat less transcendental success he also attempted to demonstrate the intelligibility of life.

Not only he succeeded in this objective, but also some of the concepts he introduced persevered in Western civilization for almost 2000 years. For example, Aristotle spoke about a cosmos whose main geometric structures were spheres. It was so difficult to eradicate this concept that scientists and philosophers, such as Digges and Bruno had to propose alternatives to the Aristotelian spheres that had even remained in the heliocentric theory of Copernicus. We shall return to this remark below.

Indeed, Aristotle proposed that the universe was formed of crystalline spheres with a common centre—the Earth, a direct influence of Anaximander. Celestial objects were attached to the various spheres each of which rotated at a fixed velocity. According to this viewpoint, the region separated by the outermost sphere (the Empyrean) was the domain of the "Prime Mover" (cf., Sec. 1.4). This was a concept introduced in the Aristotelian cosmology that was capable of triggering the initial, eternal rotation at constant angular velocity of the outermost crystalline sphere. This motion propagated from sphere to sphere, hence inducing the whole cosmos to rotate. By assuming that all the of the spheres were different and that their angular velocities were to be retrieved from observation, planetary motion became intelligible.

However, the observation that the moving planets grow dimmer or brighter could not be understood in a universe in which the spheres with their fixed celestial bodies moved with constant angular velocity, evidently at the same distance from the Earth since the crystalline Aristotelian spheres were of a fixed radius. Another mystery in Aristotle's cosmos was "retrograde motion". This remarkable observable phenomenon on the celestial sphere is as follows: as a planet moves along its orbit it gives the impression to slow down and "retrograde" (i.e., to reverse its motion), before it continues along its orbital path. Solutions to these difficulties had to wait some 400 years for the Ptolemaic system to be proposed.

Progress continued in ancient Greece: two centuries after Aristotle the Greek astronomer and mathematician Hipparchus, who died after 127 BC, anticipated Ptolemy by moving forward the frontier of astronomy. Hipparchus discovered the precession of the equinoxes. He also calculated the length of the year with some accuracy and is also remembered by his compilation of the first star catalogue. The astronomical compendium "Almagest" appeared almost three centuries after the work of Hipparchus, which very likely influenced the Ptolemaic system. Hipparchus did not adopt the views of the universe that had been proposed two centuries earlier by Aristarchus. Instead, the direct influence on the Ptolemaic geocentric system came form Hipparchus. Ptolomey's system was to prevail for over a millennium in Western civilization.

5.3 Intelligibility of the cosmos according to Ptolemy

From the point of view of astronomy, ancient culture had a climax with Ptolemy (active in Alexandria two centuries after Hipparchus, from AD 127-145). Ptolemy was the most influential astronomer of the first millennium. He adopted the view of Anaximander that the Earth was at the centre of the universe.

In order to overcome the two observational difficulties of the Aristotelian cosmology: varying brightness and retrograde motion, Ptolemy assumed that planets were attached, to circles that were in turn attached to the concentric spheres. These circles were called "epicycles", and the concentric spheres to which they were attached were termed the "deferent". Then, the centers of the epicycles described uniform circular motion as they moved at uniform angular velocity around the deferent. This fairly ingenious scheme was completed with the assumption that the planetary epicycles executed their own uniform circular motion. All speeds were constant. When even these assumptions of the Ptolemaic cosmology were insufficient. Ptolemy went on to assume that the Earth was in an eccentric position relative to the orbit. In other words the Earth was not at the centre of the circular orbit. This contributed to fitting the observational data, but he was forced to introduce another concept to understand the relative variable velocities of the planets with respect to the Earth, but maintaining the absolute motion uniform.

Retrograde motion is the consequence of this early cosmology. A further bonus yielded by this approach is that we have an implicit varying brightness. From the point of view of an Earth-bound observer, a planet appears to retrograde on the celestial sphere with distances from the Earth that also vary with time. The rather complex geometrical explanation of Ptolemy for the retrograde motion was very successful prevailing in Western culture for over a millennium.

5.4 Intelligibility of the cosmos in the Middle Ages

The Arab civilization adopted ancient culture with enthusiasm after the 7th century by who were inspired by a new monotheistic religion, Islam. In *Ideals and Realities*[3], Abdus Salam examines the reasons for the emergence of the Arab culture. Encouraged by their newly adopted monotheistic religion, they had made it their task to master science. They went on to found institutes of advanced studies. As a consequence of his effort the flourishing of their culture lasted till1450 AD with the fall of Constantinople. Their influence expanded well beyond the Arabian Peninsula. The invading Arabs appreciated and eventually reworked the classical science of the Greeks that had been carefully preserved by the early clerics in their monasteries.

In the sacred book of Islam, the Holy Koran, medicine and astronomy were seen with high esteem. So were the study of arithmetic and geometry. By 900 AD the writings of the Greeks had been translated, and thereby much of classical science had been adopted and, more significantly it was preserved, including texts on medicine, astronomy, mathematics, as well as the philosophical works of Plato and Aristotle. Ibn-Rushid, a Muslim philosopher and scientist, was also known as Averroes (while working in Cordoba for some 26 years in a library that contained over 400,000 volumes), was responsible for the translation of the works of Aristotle. He began his work in the year 1169 AD. His influence led to a renaissance in Europe during his lifetime (12th century). It is not an understatement to say that by reintroducing classical philosophy Averroes provided a basis for a discussion of the intelligibility of the cosmos amongst his contemporaries.

Beyond the creation of algebra the flourishing of science can be illustrated with their achievements in astronomy. The construction of great observatories was a cornerstone of the progress of science. It provided accurate observations for testing the Ptolemaic predictions.

Abd Al-Rahman Al-Sufi pushed the work of Ptolemy forward in his classic *"Book of Constellations"*[4]. Amongst other achievements in 1424 a major observatory was established in Central Asia at Samarkand (Uzbekistan). The most distinguished astronomer involved in this project was Ulugh Beg (1394-1449). His family had a noble background. Beg was a Mongol prince of Tartar descent. His most remarkable achievement was the development a very large sextant, namely an instrument for determining the angle between the horizon and a celestial body such as the Sun, the Moon, or a star. This instrument was used in celestial navigation to determine latitude and longitude. With this instrument Beg was able to find out exact details of planetary movements. His star catalogue had more than a thousand entries, a considerable achievement when it is seeing against the work of the Greek astronomers mentioned earlier.

By the Middle Ages, the Ptolemaic system became more significant in the West than it had previously been. Philosophy became intimately linked with the theology of Christians most significantly expressed in the writings of Thomas Aquinas. The Prime Mover became the God of natural theology; the outermost sphere of the Aristotelian cosmology became identified with Heaven. The ideas of ancient Greek philosophers were incorporated into Christianity turning them into religious dogma. In so doing the work of Thomas Aquinas (1225-1274) was of central importance. He studied under Albertus Magnus. His long association with his teacher made him a scholar deeply concerned with the Aristotelian method. He began teaching in Paris in 1257. The greatest work of Thomas was the *"Summa"* and it is the fullest presentation of his views. In Part I Thomas treats of God in much the same way as Aristotle had treated the Prime Force. God is the first cause, himself uncaused without corporeality. He suggests a fivefold proof for the existence of God that is considered as a rational designer. God governs the world as the universal first cause. In Part II Thomas develops his system of ethics, which has its root in Aristotle. He analyses the role of human reason in ethics. Part III is devoted to Christ and the question of incarnation. Christ as head of humanity imparts perfection and virtue to his members and his whole life, suffering serve this end.

5.5 The Copernican revolution, Digges and Bruno

Cosmogony concerns itself with the account of the origin of the universe. As such it is as ancient as man himself. While in classical Greek philosophy Thales of Miletus and Plato made some attempts, the earliest significant scientific accounts were developed within the gravitational theory of Sir Isaac Newton, notably by Newton himself, and later by Emmanuel Kant. The contemporary cosmogonical account has been tested in space by probes of the main space agencies of Europe and the United States that have successfully tested the basis of Einstein's theory of gravitation (General Relativity, GR).

The heliocentric theory of Nicholas Copernicus was published posthumously in 1543. During his stay at the University of Padua from 1501 to 1503, Copernicus had been influenced by the sense of dissatisfaction of the Paduan instructors with the systems of both Aristotle and Ptolemy. The new century was a time in which scientists were requiring a sense of simplicity that classical philosophy could no longer provide. The main work of Copernicus, "*On the revolution of the heavenly spheres*", was published a few months before his own death in 1543. In this influential work Copernicus placed the Sun at the center of the Solar System and inserted both the Sun and planets inside a sphere of fixed stars. Copernicus deliberately appeared to accept the tenets of Aristotle's universe, as proposed in his book, a cosmos inserted within a system of revolving planets, surrounded by a sphere of fixed stars. There is ample evidence that the Polish scientist knew of the heliocentric hypothesis of Aristarchus. Indeed, Bertrand Russell argues that the almost forgotten hypothesis of the Ionian philosopher did encourage Copernicus, by finding ancient authority for his innovation.

With Thomas Digges (1546-1595), the English mathematician and astronomer we are in the presence of a major step forward in the ascent of man towards a true understanding of his position in the universe[5]. His main contribution to the question of the plurality of worlds was to depart from the Aristotelian model of fixed spheres of stars. He translated part of *"On the revolution of the heavenly spheres"* and introduced the idea of an infinite universe with the stars at varying distances in an infinite space. From Digges work we can anticipate the clearer statement later made by Bruno of a universe in which a multitude of worlds were rotating around stars not unlike our own Sun. Copernicus did not anticipate this concept.

With the Italian philosopher, astronomer, and mathematician Giordano Bruno, (1548-1600) we reach the closing of a cycle in the gradual maturing of the science of cosmology that had begun with Anaximander 2000 years earlier[6]. Bruno is best remembered for intuitively going beyond the heliocentric theory of Copernicus, which still maintained a finite universe with a sphere of fixed stars. From the point of view of astrobiology his anticipation of the multiplicity of worlds has been amply confirmed since 1995. In that year the first detection of extra-solar planets was announced. But what is more significant regarding Bruno's intuition is that he also conjectured that such worlds would be inhabited by living beings, suggesting the possibility of a second Genesis in a modern cosmological context. His ideas went much further than those of Digges.

Astrobiology, the science of life in the universe is just concerned with this key question, still without a convincing answer. Bruno was directly influenced by the philosophy of Cardinal Nicholas of Cusa (Cusanus, 1401-1464). This Italian cardinal flourished in the middle of the 15th century. In 1440 he published a significant work: *"De Docta Ignorantia"* ("On Learned Ignorance"). In this book Cusanus denied the one infinite universe centered on Earth. He also claimed that the Sun was made of the same elements as Earth. He spoke of "a universe without circumference or center". The work of Cusanus touched on a theological question. In his system, all celestial bodies are suns representing to the same extent the explication of Divine action. Such dialogue between a scientific question (cosmology) and theology (Divine action) would lead in the subsequent century, to a conjecture that anticipated the possible existence of a plurality of inhabited worlds. It can be said that one of Bruno's major innovation, anticipated by Digges, was his refusal to accept that the Solar System is contained in a cosmos that is bounded by a finite sphere of fixed stars.

To sum up, Bruno proposes an infinite cosmos, populated by an infinite number of worlds. This proposal was first outlined in his first Oxford Dialogue: *"The Ash Wednesday Supper"*. Altogether Bruno wrote three Italian dialogues, which are relevant to our discussion. His work took place during a visit to England in 1584. These writings were in fact stimulated by controversial debates at the University of Oxford. In his third dialogue Bruno developed valuable concepts first introduced in The Ash Wednesday Supper[6]. His cosmological vision matured in Bruno's writings long before the science of astrometry allowed the concept of a multitude of solar systems to be brought within the scientific domain at the very closing of the 20th century.

5.6 Cosmogonical and cosmological models

In spite of the already mentioned early attempts to conjecture a cosmogonical hypothesis, the most influential early cosmogony in Western culture can be traced to the Book of Genesis (Gen 1:1):

"In the beginning God created the heavens and the earth".

To approach this problem from the point of view of science, a preliminary step is to grasp the significance of the scale of time involved. For this purpose we must return to the first instants of cosmic expansion[7]. In 1929 the American scientist Edwin Hubble (1889-1953) discovered that a large group of galaxies were moving away from us and that the velocity of recession of such galaxies is proportional to their distance from us. (This is known as Hubble's Law.) Light beams, as they propagate through the cosmos, have their wavelengths gradually stretched. A distant galaxy will be receding according to the Hubble Law. The observer at a certain wavelength will detect the light emitted by the galaxy. Due to universal expansion such a light beam will have a corresponding wavelength that is longer than its value when it was emitted. The expression "the light has been shifted towards the red-end of the spectrum" is an appropriate summary of this phenomenon. We say that light has been "red-shifted".

We can study cosmology in the context of a gravitational theory. Newton's theory gives reasonable overall results, but it is preferable to work in terms of the theory of gravitation, already referred to above as GR formulated by Einstein during the early part of last century (1916). It is more accurate than the Newtonian theory. In terms of this approach cosmological models may be discussed in terms of a single function R of time t. This function may be referred to quite appropriately, as a "scale factor", a measure of the size of the universe[2].

Sometimes, when referring to the particular solution, the expression "radius of the universe" R may be preferred for the scale function. The universal expansion discovered by Hubble is expressed as an increase of the radius of the universe as time t increases. The standard cosmological model assumes homogeneity in the distribution of matter (the "substratum"), as well as isotropy of space. The functional dependence of R, as a function of time t, is a smooth increasing function for a specific choice of two free parameters, which have a deep meaning in the GR theory of gravitation, namely, the curvature of space and the cosmological constant.

In 1922 the Russian theoretical physicist Alexander Friedmann (1888-1925) found the functional behavior of the scale factor R. This solution is also attributed to Howard Percy Robertson (1903-1961) and Arthur Geoffrey Walker (1909-2001) for their work done in the 1930s. Such a (standard) model is referred to as the Friedmann model. In fact, the radius of the universe R is inversely proportional to the substratum temperature T.

Hence, in the context of the expansion of the universe we have been discussing, since R is also found to increase with time t (cf., the previous paragraph), T decreases; this model implies, therefore, that as t tends to zero (the "zero" of time) the value of the temperature T is large. It should be underlined that Friedmann's model anticipated Hubble's law before it had been formulated on the basis of observational astronomy. In other words, the Friedman solution suggests that there was a hot initial condition. As the function R represents a scale of the universe (in the sense we have just explained), the expression "Big Bang", due to Sir Fred Hoyle, has been adopted for the beginning of the universe in the Friedman model. An underlying assumption is that there may be the possibility of positively curved spaces ("closed" universes), as well as negatively curved spaces "open" universes). The acceptance of the Big Bang as our present cosmology is due to its observational support. Some improvements have followed in more recent times.

5.7 The cosmic microwave background

The Big Bang model tells us that as time t increases, the universe cools down to a certain temperature, which at present is close to 3 K (three degrees Kelvin, which is equivalent to a temperature of - 270 degrees Celsius). This discovery took place in 1964 by Arno Penzias and Robert Wilson; they provided solid evidence that the part of the universe surrounding us is presently illuminated by "3 K radiation", but since it has a typical wavelength of about 2 mm due to the enormous red shift it has suffered since the moment it was last scattered during the first moments of expansion, it is referred to as the "cosmic microwave background" (CMB). In fact, we should review our understanding of what happened immediately before the formation of the CMB: in less than one million years after the beginning of the general expansion, the temperature T was already sufficiently low for electrons and protons to be able to form hydrogen atoms. Up to that moment these elementary particles were too energetic to allow atoms to be formed.

Once "recombination" of electrons and protons was possible, due to falling temperatures, thermal motion was no longer able to prevent the electromagnetic interaction from forming hydrogen atoms. This is the "moment of decoupling" of matter and radiation. At this stage of universal expansion the force of gravity was able to induce the hydrogen gas to coalesce into stars and galaxies. Hans Bethe proposed a series of nuclear reactions in the interior of stars. His aim was to understand the nuclear reactions that are responsible for nucleosynthesis. The underlying phenomenon consists of high-energy collisions between atomic nuclei and elementary particles that have been stripped off their corresponding atoms, or even nuclei, due to the presence of the enormous thermal energy in the core of the star. At such high temperatures nuclear fusion may occur. In other words, there can occur nuclear reactions between light atomic nuclei with the release of energy. In the interior of stars reactions are called thermonuclear when they involve nuclear fusions, in which the reacting bodies have sufficient (kinetic) energy to initiate and sustain the process.

One example is provided by a series of nuclear reactions that induce hydrogen nuclei (essentially single "elementary" particles called protons) to fuse into helium nuclei. A helium nucleus is heavier than the proton; in fact, it consists of two protons and two more massive particles called neutrons. This process, in addition, releases other particles and some energy. After a long phase (measured in millions of years) dominated by such conversion, or "burning" of hydrogen into helium, the star evolves: its structure becomes gradually that of a small core, where helium accumulates. To maintain the pressure balance with the gravitational force, both temperature and density of the core increase. The star increases in size. It becomes what is normally called a "red giant", because at that stage of their evolution they are changed into a state of high luminosity and red color. During this long process, from a young star to a red giant, the elements carbon and oxygen, and several others, are formed by fusing helium atoms.

The CMB may be confidently considered to be a cooled remnant from the hot early phases of the universe. It has an "isotropic" distribution, in other words, its temperature does not vary appreciably independent of the direction in which we are observing the celestial sphere (the accuracy of this statement is 10 parts per million, ppm). The isotropy is a consequence firstly of the uniformity of cosmic expansion, secondly, of its homogeneity when its age was 300,000 years and temperature of 3000 K[7]. On the other hand, in 1992 more precise measurements of the $T = 3$ K radiation, began to be made by means of the satellite called the Cosmic Background Explorer (COBE): when the accuracy of the isotropy was tested with more refined measurements, it was found that there was some degree of anisotropy after all—the temperature did vary according to the direction of observation (one part in 100,000). This fact is interpreted as evidence of variations in the primordial plasma, a first step in the evolution of galaxies.

Further accuracy in understanding the deviations from isotropy of the CMB (and hence a better understanding of the early universe) can be expected in the next few years. For example, NASA's Wilkinson Microwave Anisotropy Probe

(WMAP) has been making accurate measurements of the CMB fluctuations since 2001, WMAP. This probe has been able to produce a new high-resolution map of microwave light emitted only 380,000 years after the Big Bang. This work appears to define our universe more precisely than in previous observations, including the above-mentioned COBE probe. The eagerly awaited results announced in the years 2003-2004 WMAP resolve several long-standing disagreements in cosmology rooted in less precise data. Specifically, present analyses indicate that the universe is 13.7 billion years old (accurate to 1 percent), composed of 73 percent dark energy, 23 percent cold dark matter, and only 4 percent atoms, is currently expanding at the rate of 71 km/sec/Mpc (accurate to 5 percent), underwent episodes of rapid expansion called inflation, and will expand forever.

5.8 From the composition of the universe to eventual insights into its destiny

While in classical Greek philosophy Thales of Miletus and Plato made attempts to understand the cosmos, the earliest scientific accounts were developed within the gravitational theory of Sir Isaac Newton, notably by Newton himself and later by and Emmanuel Kant. The contemporary cosmogonical account has been tested in space by probes of the main space agencies of Europe and the United States. They have successfully tested the bases of Albert Einstein's theory of gravitation.

From the previous section we have seen how measurements of the CMB yield some information on the earliest stages of the universe. Our present understanding of the composition of the universe in terms of dark matter and dark energy, on the other hand (as stated in Sec. 5.7), remains a challenge for the future.

Firstly, our insights into dark matter have arisen as follows. All visible and microscopic form of matter is made up of protons, neutrons and electrons. Protons and neutrons are bound together into nuclei. Atoms are nuclei surrounded by electrons. For instance, hydrogen is composed of one proton and one electron. Helium is composed of two protons, two neutrons and two electrons. Carbon is composed of six protons, six neutrons and six electrons. Heavier elements, such as sulfur, iron, lead and uranium, contain even larger numbers of protons, neutrons and electrons. The universe is not composed entirely of this "baryonic matter". There is evidence that suggests there is an additional form of matter that has been called "dark matter"[8]. This concept arose thanks to the insights of Fritz Zwicky (1898-1974). He was a Bulgarian astrophysicist, who held a professorial position at the California Institute of Technology from 1925 till 1968. He was responsible for introducing the concept of missing mass. In collaboration with Walter Baade he based his observations on the Coma Cluster that is some 300 million light years away. This is one of two clusters that lie in the constellation of Coma Berenices (the other is the Virgo Cluster). It is a remarkable cluster as it contains over a thousand galaxies with an average distance between them that is about three times smaller than the distance between the Milky Way and Andromeda. Zwicky was the first to realize that there should be a great deal of invisible matter to make the cluster stable. Such invisible matter is conjectured to be one form of what we now call dark matter. Indeed, the mass inferred for galaxies, including our own, is roughly ten times larger than the mass that can be associated with stars, gas and dust in a Galaxy. This mass discrepancy has been confirmed by observations of the effect of gravitational lens, the bending of light predicted by the Theory of General Relativity. By measuring how the foreground cluster distorts the background galaxies, we can measure the mass in the cluster. The mass in the cluster is more than five times larger than the inferred mass in visible stars, gas and dust (baryonic matter). The remaining mass contribution of the cluster arises from matter that, as we mentioned above, we call dark matter that is matter exerting a gravitational attraction, but does not emit nor absorb light, as can be inferred form the Hubble Space Telescope images from which these studies have been conducted. For theoretical reasons some candidates for dark matter have been suggested: brown dwarfs, supermassive black holes and even new forms of matter.

Secondly, dark energy is still another challenge[9]. In 1998 two groups studying supernovae showed that a fraction of energy of the cosmos is accelerating the expansion of the Universe. Subsequent work revealed that dark energy may make up about 70 percent of the Universe, but our current understanding of cosmology based on the General Theory of Relativity is unable to explain this unknown form of energy. Our hope for progress is based on the many missions that the major space agencies are planning. For example, a more accurate gravitational theory may emerge form the Laser Interferometer Space Antenna (LISA). This is jointly sponsored by ESA, as a Cornerstone mission in their "Cosmic Vision Program", and NASA's Structure and Evolution of the Universe 2003 Roadmap, "Beyond Einstein: From the Big Bang to Black Holes." LISA will test the Theory of General Relativity, probe the early Universe, and last but not least, LISA will search for gravitational waves. LISA is a space-based gravitational-wave observatory, capable of detecting waves generated by binaries within the Milky Way, and by waves generated by massive black holes in other galaxies. LISA will use a system of laser interferometry by directly detecting and measuring gravitational waves.

In addition the instrumentation form the physics of the high energies may also offer further help in improving our understanding of the universe composition and hence its eventual destiny: About 100 m below us, in a tunnel that runs in a ring for 27 km when it is switched on in the summer 2007, this experiment will collide two beams of particles head-on recreating the conditions in the Universe moments after the Big Bang creating showers of new particles that will reveal new physics beyond the Standard Model. The main aspects of the Large Hadron Collider (LHC) experiment are the LHC detectors and the LHC particle beams.

They are named LHCb, Alice, Atlas and the Compact Muon Solenoid (CMS). While LHCb and Alice are designed to investigate specific physical phenomena, Atlas and CMS are designated "general purpose" detectors. They will both aim to identify the elusive Higgs boson important for the validity of the Standard Model. This theory is a major achievement of physics of the 20th century. Its microscopic description incorporates the interactions of three sub-atomic forces: they include the force that holds the atomic nucleus together, the force responsible for nuclear beta decay and electromagnetism. With the progress of observational cosmology, we now realize that the Standard Model cannot incorporate gravity. It is also restricted to ordinary matter, which makes up only a small part of the Universe, excluding implicitly dark matter. The LHC aims to add a missing ingredient of the Standard Model: A hypothetical subnuclear particle that explains why all other particles have mass. According to the Standard Model, particles acquire their mass by means of a mechanisms theorized by Peter Higgs in 1964. According the Standard Model, there is a mechanism that explains how two types of particles, massless like everything else immediately after the Big Bang, came to acquire different masses as the universe cooled down. The Higgs particle could shed light on the still-to-be understood nature of dark energy. With the help of Atlas and CMS will search for new kinds of particles. New forces and new types of particles may be the constituents of dark matter. One of the primary motivations for LHC is to try to produce this matter in Geneva. Since the universe was very dense and hot in the early moments following the Big Bang, the universe itself was a "particle accelerator". Dark matter may be made of weakly interacting massive particles that were produced shortly after the Big Bang.

The instrumentation that has been developed for extending our grasp of the nature and intelligibility of the universe is impressive. For example, the particle detector of the LHC, Atlas, is 46 m long and is 25 m high. It weighs 7000 tons. The instrumentation is even more impressive by its use of operating temperature of 1.9 Kelvin (-271 C). This is just above "absolute zero" and colder than the vacuum of outer space. The magnets are cooled to this ultra-low temperature by bathing them in liquid helium, so as to take advantage of phenomena typical of the very low temperatures that make helium ideal for cooling and stabilizing a large system such as the instrumentation of the LHC.

Chapter 6

The Destiny of Life in the Universe

We outlined in Chapter 5 some stepping-stones towards the intelligibility of our nearby cosmic environment, where life is known to have emerged on our planet. In this and subsequent chapters we discuss why the concept of convergence in the life sciences has to be paid some attention in any well-focused pursuit of the intelligibility of nature. In our search for understanding intelligibility we pursue the dialogue of science with natural theologians, who, unlike scientists, are concerned with questions of purpose.

The reader is advised to refer especially to the following entries in the Glossary: Archaean, Archaea, Carbonaceous chondrite, Eukaryogenesis, Eukaryote, Extremophile, Nucleosynthesis, Oro, Phylogenetic tree, RNA, Supernovae.

6.1 Cosmic pathways towards a second Genesis

Stars whose mass is similar to that of the Sun remain at the red giant stage for a few hundred million years[1]. The last stages of burning produce an interesting anomaly: the star pushes off its outer layers forming a large shell of gas; in fact, the shell is much larger than the star itself. This structure is called a planetary nebula. The star itself collapses under its own gravity compressing its matter to a degenerate state, in which the laws of microscopic physics (called quantum mechanics) eventually stabilize the collapse. This is the stage of stellar evolution called a "white dwarf". After the massive star has burnt out its nuclear fuel (in the previous process of nucleosynthesis of most of the lighter elements), a catastrophic explosion follows, in which an enormous amount of energy and matter is released. It is precisely these "supernovae" explosions that are the source of enrichment of the chemical composition of the interstellar medium. This chemical phenomenon, in turn, provides new raw material for subsequent generations of star formation.

Late in stellar evolution stars are still poor in some of the heavier biogenic elements (such as, for instance, magnesium and phosphorus). Such elements are the product of nucleosynthesis triggered in the extreme physical conditions that are due to the supernova event itself. By this means the newly synthesized elements are disseminated into interstellar space, becoming dust particles after a few generations of stars births and deaths.

As a product of several generations of stellar evolution, a large fraction of the gas within our galaxy is found in the form of clouds of molecular hydrogen, but also many of the heavier elements are present too. The linear dimension of these clouds can be as much as several hundred light years. The mass involved may be something in the range 10^5 to 10^6 solar masses. Images from the orbital Hubble Space Telescope (HST) have shown circumstellar disks surrounding young stars. This supports the old theory of planetary formation from a primeval nebula surrounding the nascent star. In the case of our solar system this gas formation has been called the "solar nebula". We have already seen that stars evolve as nuclear reactions convert mass to energy. In fact, stars such as our Sun follow a well-known pathway (the main sequence) along a Hertzprung-Russell (HR) diagram. Ejnar Hertzprung and Henry Russell introduced this diagram independently. They observed that many nearby stars have analogous physical properties: this similarity is evident when we plot luminosity (the total energy of visual light radiated by the star per second) versus its surface temperature or, alternatively, its spectroscopic type. Such stars are called main sequence stars.

We may ask: *How do stars move on the HR diagram as hydrogen is burnt?* Extensive calculations show that main sequence stars are funneled into the upper right hand of the HR diagram, where we find red giants of radii that may be 10 to 100 times the Sun radius. Stellar evolution puts a significant constraint on the future of life on Earth, since the radius of the Earth orbit is small.

6.2 Astrochemistry

The chemistry of the circumstellar zones sets the stage for prebiotic evolution. In other words, a few molecules are sufficient for the synthesis of the key biomolecules of life, namely, the amino acids, the bases, sugars. By looking at other solar systems in the process of formation, we are led to conjecture what was the nature of our own solar system. It is reasonable to assume that the solar system was formed out of a disk-shaped cloud of gas and dust, which we called the solar nebula.

Most of the original matter of the solar nebula has since been incorporated into its planets and the central star itself, namely, the Sun. The nature of the interstellar dust may be appreciated as the product of condensation of "metallic elements" We adopt the terminology of astronomers, in which any chemical element with mass higher than hydrogen or helium is called "metallic".

Metallic elements were themselves produced in stellar interiors (for example, magnesium, silicon and iron). Later, in interstellar space they combined with elements such as oxygen to form particles measuring typically 0.1 microns. A very significant clue regarding the nature of the original solar nebula can still be inferred from the study of comets. In fact, since these small bodies remained far from the star, its chemical evolution was insignificant, and consequently they remained essentially unchanged. We have to wait for the Rosetta mission to retrieve a piece of a comet in the second decade of this century. Yet, as comets pass in the vicinity of the Sun and the Earth, dust particles are released as the comets heats up. NASA's Stardust mission had access to verified samples of a comet[2]. The leftover debris from the formation of the Solar System 4.5 billion years ago, comets consist mostly of ice, dust and rock. The Stardust mission was launched in February 1999. It carried a set of instruments to monitor the impact of cometary dust. On January 2, 2004, the spacecraft came within 150 miles of the comet and collected thousands of tiny dust particles streaming from its nucleus. The Stardust sample-return canister parachuted onto the desert salt flats of Utah the following year. The comet dust is now available for comparisons to tiny particles that are constantly raining down on Earth that scientists suspect come from comets.

Such interplanetary dust particles (IDPs) are a major component of our galaxy, including the Solar System[3,4]. In the early 1960s John Oro was responsible for the conjecture that most of the volatiles on Earth (substances with low boiling point) had been delivered by comets-the most striking example being the water of the oceans.

Given the importance of the origin and evolution of the biomolecules, we feel that it is appropriate to comment on the origin of some of the most important biomolecules. They may have been synthesized in the early Earth, or transported here from space. There are also many indications that the origin of life on Earth may not exclude a strong component of extraterrestrial inventories of the precursor molecules that gave rise to the major biomolecules. About 98 percent of all matter in the universe is made of hydrogen and helium. Besides hydrogen, other biogenic chemical elements, namely, C, N, O, S and P, make up about 1 percent of the cosmic matter. Some of the materials having a higher melting point are called "refractory". They have gone through chemical reactions at higher temperatures. Consequently, refractory materials have remained in the inner solar and have condensed in the form of meteorites. These small bodies are called "carbonaceous chondrites", which are thus testimonies of the nature of the solar nebula. They are of special interest as they also contain some of the above-mentioned biogenic elements, whose abundance would suggest that the major part of the molecules in the universe would be based on carbon (normally they are referred to as being "organic molecules"). In fact, out of over a hundred molecules that have been detected, either by microwave or infrared spectroscopy, 75 percent are organic.

Once again, chemical evolution experiments fare well in comparison with the observation of the interstellar medium. Some of the identified molecular species detected by means of radio astronomy are precisely the same as those shown in the laboratory to be precursor biomolecules.

6.3 The origin and evolution of the Solar System

Before reviewing chemical evolution itself, we should first understand current ideas of how the Earth itself originated. The origin of the solar system goes back to about 4.5 Gyr BP by the gravitational collapse of the solar nebula when a certain critical mass was reached.

Thus, the protosun, the protoplanets, the comets, the parent bodies of meteorites, and other planetesimal bodies were formed as the result of this condensation of interstellar matter. The lunar cratering record demonstrates that during its initial stages, the solar system may have been in a chaotic state with frequent collisions of planetesimals amongst themselves as well as with other larger bodies, including the protoplanets. The composition of the solar nebula must have been analogous to that of the interstellar clouds, namely it may have consisted of hydrogen, helium, as well as carbon compounds. This has been confirmed by astronomical studies of Jupiter, Saturn and their satellites.

To get a deeper insight into the origins of the solar system we need to consider its small objects. We will restrict ourselves to comparisons between the densities of carbonaceous chondrites and the rocky planets, or satellites; we know from such comparisons that the terrestrial planets were not formed by a slow process of gradual accumulation of interstellar dust particles that may have fallen in the solar nebula. Instead, the outline of what may have separated the constituent globules ("chondrules") observed in the carbonaceous chondrites from gases in the nebula is based on considerations of chondrites as relics of the accretion episode. From the combined information provided by the small objects of the inner solar system (chondrites) and messengers form the outer solar system (comets), we can appreciate an outline of the main steps in the formation of the terrestrial planets. We expect heat at the centre of the solar nebula and low temperatures at its periphery. This is today an empirical observation for we can observe in our galaxy nebulae where similar processes of star formation are occurring today.

As the thin interstellar medium gets concentrated in some regions in space it forms a protonebula that eventually collapses onto itself, thereby producing a hot interior. The collapse is coupled with rotation of the gas. Further out from the inner region in the solar nebula that will eventually give rise to the terrestrial planets, gas is more abundant than IDPs, and cool enough to allow the existence of water ice, which would be more abundant by volume than the dust grains. Without going very deep into the details, we can already appreciate why the terrestrial planets are denser than the Jovian planets. Besides, since at a distance of 5-10 AU the temperature was not low enough to discriminate gases form dust, the composition of Jupiter and Saturn that were formed at these distances, are expected to reflect the composition of the solar nebula.

6.4 From chemical to cellular evolution

The major steps towards life should have taken place from 4.6-3.9 Gyr BP, the preliminary interval of geologic time that is known as the "Hadean Subera". It should be noticed that impacts by large asteroids on the early Earth do not necessarily exclude the possibility that the period of chemical evolution may have been considerably shorter.

Indeed, it should not be ruled out that the Earth might have been continuously habitable by non-photosynthetic ecosystems from a very remote date, possibly over 4 Gyr BP. The content and the ratios of the two long-lived isotopes of reduced organic carbon in some of the earliest sediments (retrieved from the Isua peninsula, Greenland, some 3.8 Gyr BP) may convey a signal of biological carbon fixation[5].

These arguments reinforce the expectation that chemical evolution may have occurred in a brief fraction of the Hadean Subera, in spite of the considerable destructive potential of large asteroid impacts that took place during the same geologic interval in all the terrestrial planets, the so-called *heavy bombardment period*[6]. In subsequent sub-eras of the Achaean (3.9-2.5 Gyr BP) life, as we know it, was present. This is represented by fossils of the domain Bacteria that is documented by many species of cyanobacteria.

6.5 The terrestrial tree of life

In spite of the fundamental work of Darwin that gave rise to modern biology, earlier in the 20th century the fact remained that the Earth biota was still being divided into animals and plants. It was only in the 1930s when taxonomy shifted its emphasis form the multicellular dominated classification to one more oriented towards basic cellular structure. The encapsulation of chromosomes in nuclei was clearly absent in bacteria. This remark led to a division of all living organisms into two groups that went beyond the animal/plant dichotomy.

However, the tree of life could not yet be constructed, since amongst bacteria it was remarked that rapid and random exchange of genes occur. This is a phenomenon called horizontal gene transfer (HGT)[7,8]. This randomness deterred the use of sequences of the biomolecules, in order to construct a tree of life that is technically known as a phylogenetic tree. Yet, Linus Pauling and Emil Zuckerkandl had pointed out that a "molecular clock" might be identified from the slow mutation rates of some biomolecules.

The question then is to identify the molecule that would not be affected by HGT, as evolution proceeds. Such molecules are a form of RNA that together with some protein constituents make up the ribosomes, the small corpuscles of the cell where proteins are synthesized. This form of RNA is called ribosomal RNA (rRNA). Extensive work with these molecular chronometers has led to a taxonomy including all life on Earth that is deeper than the simple dichotomy prokaryote/eukaryote. A classification in which the highest taxa are called domains, instead of Kingdoms, was put forward. In this approach there are three "branches" in the tree of life[9]:

- Archaea, whose cellular membrane differs from the structure of the lipid bilayer of Bacteria and Eucarya, but like Bacteria they do not possess an internal nucleus. Using a word of Greek origin for nucleus (karyon), Archaea and Bacteria are called "prokaryotes".
- Bacteria, encompassing all bacteria.
- Eucarya, including all the truly nucleated cells.

Many "extremophilic" microorganisms are known, namely microorganisms that are adapted to extreme environments, such as heat, high salinity, or unusual chemical environments. For example, 'hyperthermophilic' archaea (i.e., extremophiles that survive at especially high temperatures) have been isolated from deep-sea hydrothermal vents.

Besides, today hydrogen sulfide and methane are abundant in the "black smokers" of the seafloor (fluids of volcanic origin). "Chemothrops" is a term applied to microorganisms that obtain the energy they require independent of photosynthesis to fix carbon dioxide, and to produce the organic matter that is needed. In such environments, the seafloor and deep underground, microorganisms thrive in the most extreme conditions. These microbes are likely candidates for the dawn of cellular life on Earth. Several lines of research suggest the absence of current values of oxygen, O_2, for a major part of the history of the Earth. Some arguments militate in favor of Archaean atmospheres with values of the partial pressure of atmospheric oxygen O_2 about 10^{-12} of the present atmospheric level (PAL). The growth of atmospheric oxygen was due to the evolutionary success of cyanobacteria that were able to extract the hydrogen they needed for their photosynthesis directly from water. One of the chief indicators of the growth of atmospheric oxygen is a rock (shale) that has played a role in our understanding of biological evolution. The onset of atmospheric oxygen is demonstrated by the presence in the geologic record of shale that has been colored by ferric oxide (and hence it is red). The age of such "red beds" is estimated to be about 2 Gyr. At that time oxygen levels may have reached 1-2 percent PAL, sufficient for the development of a moderate ozone (O_2) protection from ultraviolet (UV) radiation for microorganisms from the Proterozoic (2.5 billion years to 570 million years before the present). In fact, UV radiation is able to split the O_2 molecule into the unstable O-atom, which, in turn, reacts with O_2 to produce O_3 (ozone) that is known to be an efficient filter for the UV radiation.

The paleontological record suggests that the origin of eukaryotes occurred earlier than 1.5 Gyr BP[10]. There are Achaean rock formations (which may be found up to 2 Gyr BP) that are significant in the evolution of life. Detailed arguments based on geological evidence supports the conclusion that we had to wait until about 2 Gyr BP for a substantial presence of free O_2. Once the eukaryotes enter the fossil record, its organization into multicellular organisms followed in a relatively short period (in a geological time scale). Inevitable random factors driving evolution of life on Earth were the mass extinctions that occurred sporadically in the past. These were due to large impacts on Earth by comets and meteorites. An example is the collision in the Yucatan Peninsula, Mexico, at the end of the Cretaceous some 65 million years BP. It is generally agreed that it was a factor in the extinction of the dinosaurs.

6.6 The evolution of life in the universe

The progress in understanding our own origins in the middle of last century, triggered by the success in retrieving some key biomolecules in experiments that attempted to simulate prebiotic conditions. Organic chemists did some of the main experiments of the 1950s and early 1960s. Since that time the field has continued its robust growth. These efforts have led to view the cosmos as a matrix in which organic matter can be inexorably self-organized by the laws of physics, chemistry and biology into what we recognize as living organisms.

However, in this context it should be stressed that chemical evolution experiments have been unable to reproduce the complete pathway from inanimate to living matter. In particular, the prebiotic synthesis of all the RNA bases is not clear and cytosine, for instance, may not be prebiotic and it may even have been imported from space. Thus, at present the physical and chemical bases of life that we have sketched are persuasive, but further research is still needed. Originally the subject began to take shape as a scientific discipline in the early 1920s, as we have seen in Sec. 2.5, when Oparin applied the scientific method of conjecture and experiments to the origin of the first cell, thereby allowing scientific enquiry to shed valuable new light on a subject that has traditionally been the focus of philosophy and theology.

Darwin's insights offer some explanation for the existence of life on Earth, and allows for meaningful questions regarding early terrestrial evolution. With the enormous scope of bacterial evolutionary data available today from an extensive micropaleontological record, Darwin's *Theory of Common Descent* leads back to a single common ancestor, a progenote or "cenancestor". The characteristics of the cenancestor can be studied today through comparison of macromolecules, allowing researchers to distinguish and recognize early events that led to the divergent genesis of the highest taxa (domains) among microorganism. Finally, at the other extreme we have the search for extraterrestrial intelligence SETI that was introduced earlier in Sec. 2.1.

6.7 Pathways towards intelligence in the cosmos

What is significant for astrobiology is to recognize that natural selection necessarily seeks solutions for adapting evolving organisms to a relatively limited number of possible environments. We have seen in astrochemistry that the elements used by the macromolecules of life are ubiquitous in the cosmos. We have seen that the formation of solar systems is limited by a set of physical phenomena, which repeat themselves in the Orion nebula; those Jovian planets have been discovered in our "cosmic village".

To sum up, the finite number of environments forces upon natural selection a limited number of options for the evolution of organisms. From these remarks we expect convergent evolution to occur repeatedly, wherever life arises. It will make sense, therefore, to search for the analogues of the attributes that we have learnt to recognize in our own particular planet. We may argue that not only life is a natural consequence of the laws of physics and chemistry, but once the living process has started, then the cellular plans, or blueprints, are also of universal validity: the lowest cellular blueprint (prokaryotic) will lead to the more complex cellular blueprint (eukaryotic). This is in principle a testable hypothesis in forthcoming space missions if we are confronted with a second Genesis.

Within a decade or two a new generation of space missions may be able to undertake significant evolutionary experiments, but alas missions that are at present in the process of study of their feasibility are still unable to face such challenges. Closely related to the proposed universality of eukaryogenesis, there are different possible conjectures regarding the question of extraterrestrial life: *Is it reasonable to search for Earth-like organisms, such as a eukaryote, or should we be looking for something totally different?*

Intelligence, as we understand it, had to await the emergence of the *Homo* genus some 2 Myr BP, when its first traces appeared in our ancestors. A more evident demonstration of the appearance of intelligent life on Earth than the tools of the habilines (*Homo habilis, Homo rudolfensis*), or the ceremonial burials had to wait until the emergence of the Magdalenian "culture". This group of human beings flourished from about 20,000 to 11,000 years before the present (yr BP). The Magdalenians left some fine works of primitive art as, for instance, the 20,000 year-old paintings on the walls of caves discovered in various places in Europe, including the caves of Altamira, as mentioned in the Preface.

The problem of communication amongst intelligent beings from different stellar environments has eventually to be faced in truly scientific terms. We consider two evolutionary pressures, terrestrial and aquatic that have given rise to the largely different brains of dolphins and humans (although some general features are common to both of them, such as the modular arrangement of neurons). The intelligence of humans and dolphins are products of a pathway that began from the general prokaryotic blueprint appearing on Earth almost instantaneously (in a geologic time scale), and continued through the sequence of eukaryogenesis, neuron, multicellularity and, finally, brains.

The present discussion of universal aspects of communication amongst humans at some point has to be extended from the single tree of life that we are familiar with on Earth. Life may not have remained confined to the species of our single tree of life (technically called a phylogenetic tree). In other words, eventually we should be in a position to discuss independent evolutionary lines. They may have evolved on planets around other stars (exoplanets), or in our own solar system. Elsewhere in this book we have touched upon the possibly of life that may have evolved on Mars, or even on the satellites of the Outer Solar System around Jupiter (Europa), or Saturn (Enceladus and Titan).

Chapter 7

Towards the Intelligibility of Nature

The first six chapters have led us half way along our search for deeper insights into the intelligibility of two aspects of nature: the evolution of the cosmos and the life that has emerged a few billion years after the Big Bang. By providing illustrations of convergent phenomena occurring at a cosmic level in this chapter we will expose our readers to some of the many phenomena beyond chemistry that need to be discussed in our efforts to understand the possibility of a second Genesis. Examples will help us to approach the question of how contemporary science, especially astrobiology, has attempted to explain the intelligibility of nature[1].

The reader is advised to refer especially to the following entries in the Glossary: Cassini-Huygens mission, Galileo mission, Interstellar dust, Pioneers, Regolith, Solar wind, Space weather, Supernova and Ulysses mission,

7.1 Convergence in the synthesis of the elements

Hydrogen and helium make up almost the totality of the chemical species of the Universe. Only 2 percent of matter is of a different nature, of which approximately one half is made by the five additional "bio-friendly", or biogenic chemical elements (C, N, O, S, P). It is clear that as a group the elements that are needed for life to prevail in the universe. In fact, carbon is the most abundant element after hydrogen and helium. So it is not surprising that the most abundant molecules detected by means of microwave spectroscopy are organic molecules.

This is an important factor that is at the basis of the fitness of the cosmos for the origin and evolution of life. We know from the previous chapter that stellar evolution that nuclear synthesis is relevant for the generation of the elements of the Periodic Table beyond hydrogen and helium.

Furthermore, nuclear synthesis is also relevant in the long term for the first appearance of life on Earth. The elements synthesized in stellar interiors are needed for making the organic compounds. This is a basic requirement. They have been observed in the circumstellar, as well as in interstellar medium, in comets and other small bodies. The same biogenic elements are also needed for the synthesis of biomolecules of life. Besides, the spontaneous generation of amino acids in the interstellar medium is suggested by general arguments based on biochemical tests.

The experimental detection of amino acids in the room-temperature residue of an interstellar ice analogue has yielded 16 amino acids, some of which are also found in meteorites[2,3]. These are factors that help us to understand the first steps in the eventual emergence of habitability on planets.

7.2 Convergence in the delivery of the biomolecules

Small bodies, such as the Murchison meteorite, may even play a role in the origin of life: according to chemical analyses in this particular meteorite we find basic molecules for the origin of life such as lipids, nucleotides, and over 70 amino acids[4]. Most of the amino acids are not relevant to life on Earth and may be unique to meteorites.

This remark demonstrates that those amino acids present in the Murchison meteorite (cf., Sec. 7.4), which also play the role of protein monomers, are indeed of extraterrestrial origin. If the presence of biomolecules on the early Earth is due in part to the bombardment of interplanetary dust particles, or by the collision with comets and meteorites, then the same phenomenon could be taking place in any other solar system. In addition, chemical analysis has exposed the presence of a variety of amino acids in the Ivuna and Orgueil meteorites as well[5].

7.3 Convergence in the formation of interstellar gas

Solar systems, many of which are known so far, originate out of "interstellar dust", that is dust constituted mainly out of the fundamental biogenic chemical elements of life. Just before a star explodes into its supernova stage, all the elements are expelled that were originated by thermonuclear reactions its interior. Typical planetary nebulae are formed at a later stage of stellar evolution from expanding clouds of dust and gas. In stellar evolution in an advanced stage the star increases in size and undergoes a change in surface temperature, responsible for a red color (a "red giant" stage). Stars with mass similar to that of the Sun remain at the red giant stage for a few 100 million years. The star itself collapses under its own gravity compressing its matter to a degenerate state, in which the laws of microscopic physics eventually stabilize the collapse: this is the stage of stellar evolution called a "white dwarf". Stellar evolution of stars more massive than the Sun is far more interesting: after the massive star has burnt out its nuclear fuel a catastrophic explosion follows, in which an enormous amount of energy and matter is released. These events were called "supernovae" explosions in Sec. 6.1. They are the source of enrichment of the chemical composition of the interstellar medium. This phenomenon, in turn, provides new raw material for subsequent generations of star formation that leads to the formation of planets.

At a late stage of stellar evolution stars are still poor in some of the heavier biogenic elements (such as, for instance, magnesium and phosphorus). Such elements are the product of nucleosynthesis triggered in the extreme physical conditions that occur in the supernova event itself. By this means the newly synthesized elements are disseminated into interstellar space, becoming dust particles after a few generations of star births and deaths.

7.4 The evolution of the Sun and the origin of life

The Solar System formed in the midst of a dense interstellar cloud of dust and gas. This event may have been triggered by the shock wave of a supernova explosion. Indeed, there is some evidence for the presence of a chemical clue: a compound (silicon carbide, SiC grains) in the famous meteorite that was already referred to in the previous section. It has been demonstrated that the source of such a compound is from a supernova (called "type II supernova"[6]). The Solar System had such a beginning over 4.6 billion years before the present (Gyr BP).

The effects of solar activity and space weather have played a leading role in the origin and evolution of life on Earth. Our insights into this subject have improved considerably in recent years, due to several factors. Firstly, we have the advent of a long series of space missions. Secondly, we remark that improved knowledge of the Sun and space weather is due to a large extent to the advent of robust national efforts, such as the Institute of Atmospheric Sciences and Climate (ISAC) of the Italian National Research Council, the Japanese Institute of Space and Astronautical Science (ISAS), and the National Aeronautics and Space Administration (NASA).

These single-nation efforts are complemented by the international collaboration of the European Space Agency (ESA). We need to understand chemical and biological evolution that took place on Earth. In addition, as our star approached the main sequence of the HR diagram (cf., Sec. 6.1), the continually evolving early Sun was providing necessary energy requirements for the emergence of life. We discuss possible favorable, and also inhibiting effects on the origin of life that were provided by the influence of solar evolution on space weather. We especially keep in mind that knowledge of the early evolution of solar radiation still presents many challenges, the solution of which are not always free of controversy.

Two major factors during this early period were the non-ionizing effects of ultraviolet radiation, as well as the ionizing effects of X- and gamma rays. By not restricting ourselves exclusively to solar effects, but rather to space weather effects (with possible inputs from galactic events), we gain a wider perspective on the importance of improved understanding of solar activity and its intimate relation with the origin of life on Earth.

7.5 Convergence in the direct records of solar evolution

An important factor for understanding the origin and evolution of life on Earth is the evolution of the Sun itself. Many aspects of its history remain to be understood. We reconsider the constraints that present knowledge of our own star implies for the emergence of life on Earth. This, in turn, will provide further insights into what possibilities there are for life to arise in any of the multiple solar systems that are known to date.

Fortunately, the particles that have been emitted by the Sun in the past have left a record in geologic samples belonging to the Hadean (4.6-3.8 billion years before the present, Gyr BP) and Archaean (3.8-2.5 Gyr BP). It is generally agreed that the latter period corresponds with the emergence of life, but we cannot exclude completely possible earlier dates for the onset of life on Earth. Considerable information can be retrieved from observations of extraterrestrial samples, either meteorites, or lunar material[6]. We may gather together a few of these constraints with further constraints arising from the cratering record of the Moon. This information may be supplemented with studies involving data related to the fractionation of the stable isotopes of hydrogen, carbon, nitrogen, sulfur and the noble gases. We remind the reader that isotopes are one or more atoms of the same chemical element that have the same number of protons in their nuclei, but different number of neutrons. The notation used is to precede the symbol of the chemical element with an index denoting the total number of particles in the nucleus. For example, the isotope of sulfur that contains 34 neutrons and protons is denoted as ^{34}S. In addition, the reader is also reminded that a set of gases are called "noble" when they are formed of the atoms of he following chemical elements: helium, neon, argon, krypton, xenon and radon.

The solar wind records in the earliest samples available to us in the small bodies of the Solar System were irradiated during the Archaean. They have been studied for their chemical elements, as well as for their isotopic composition, taking advantage of the fact that meaningful comparisons can be made with current observations of the solar wind[7].

A very early origin of life has to take into account the imprints of such solar energetic particles during the first Gyr after the formation of the Sun. There is some evidence for a more active Sun in the past. For instance, we are aware of an excess of the isotope ^{21}Ne in grains from the Murchison meteorite.

7.6 Isotopic fractionation of noble gases on Earth

Chemical or physical processes that act to separate isotopes provide an important signature of the early Sun. These processes are known as producers of an "isotopic fractionation". A particularly important case is the isotopic fractionation of the five stable noble gas elements. The early atmosphere arose from collisions during the accretion period, the so-called heavy bombardment period of the surface of the Earth during the early Archaean (before 3.8 Gyr BP). Planetesimal impacts increase the surface temperature affecting the formation of either a proto-atmosphere or a proto-hydrosphere by degassing of volatiles[8]. This generated a "steam atmosphere". One of its consequences was a rapid hydrodynamic outflow of hydrogen, including some of its compounds carrying along heavier gases in its trail [9]. The mechanism postulated is that of aerodynamic drag. The upward drag of noble gas atoms of similar dimension competes with an opposite force due to gravity. Hence, since the various isotopes of these gases have different masses the result is the occurrence of a mass-dependent fractionation of the noble gas isotopes.

For even heavier atoms, the gravity effect can be stronger than the aerodynamic drag and such atoms would not show the remarkable fractionation typical of the noble gases. By looking at other main-sequence stars at equivalent early periods of their evolution, we became aware of an associated larger output of solar radiation in the wavelength range 100-120 nm, the so-called extreme ultraviolet radiation. With the early Sun such an ultraviolet excess radiation is a possible factor that can trigger the phenomenon of mass fractionation in the noble gases. The case of the $^{22}Ne/^{20}Ne$ ratio is an example, since its value is larger than in the Earth's mantle, or in the solar wind. The observed fractionation of the noble gases can be taken as a signature of two aspects of the early Sun: firstly, the presence of the postulated escape flux, and secondly, as evidence for the solar energy source that drives the outward flux of gases. The emergence of life on Earth has to wait until the decrease of solar radiation that characterizes the terrestrial accretion period.

The beginning of such a favorable period begins once accretion has ended. The surface heat flux diminishes, leading to the steam atmosphere raining into a global ocean[10]. This splitting of a primitive atmosphere into a hydrosphere and a secondary atmosphere leaves behind carbon and nitrogen compounds that will be ingredients for subsequent steps of chemical evolution and the dawn of life.

7.7 The solar wind after the Genesis mission

The expansion of the solar corona induces a flux of protons, electrons and nuclei of heavier elements (including the noble gases) that are accelerated by the high temperatures of the solar corona, or outer region of the Sun, to high velocities that allow them to escape from the Sun's gravitational field. The solar wind is so tenuous that at a distance of 1 AU during a relatively quiet period, the wind contains approximately five particles per cubic centimeter moving outward from the Sun at velocities of 3×10^5 to 1×10^6 ms^{-1}; this creates a positive ion flux of just over 100 ions per square centimeter per second, each ion having an energy equal to at least 15 electron volts. The solar wind reaches the surface of the Moon modifying considerably its upper surface or "regolith".

We have considerable information on the lunar regolith thanks to the Apollo missions in the years 1969-1972 that retrieved so much material and made it available to many laboratories, including those of John Oro and Cyril Ponnamperuma that influenced so much of our early understanding of the origin of life on Earth, as we have explained elsewhere in this book (cf., Chapter 3). By so doing the solar wind modifies its structure leaving a telltale hint of how it changes over geologic time, since the Moon is an inactive body only being modified by the impacts of external objects such as meteorites.

Much more recently, the Genesis mission was NASA's first sample return mission sent to space. It was the fifth of NASA's Discovery missions. Genesis was launched in the year 2001 with the intention to bring back samples from the Sun itself. Three years later, after crash-landing, the probe was retrieved in Utah. Genesis collected particles of the solar wind on wafers of gold, sapphire, silicon and diamond. The amount of stardust collected by Genesis was about 10^{20} ions, or equivalently, 0.5 milligrams. Preliminary studies indicate that contamination did not occur to a significant extent. Genesis is providing us with important information for getting a deeper understanding of the early Solar System, and hence a better opportunity for a closer approach to the mystery of the origin of life on Earth.

The Moon is depleted of volatile elements such as hydrogen, carbon, nitrogen and the noble gases, possibly due to the fact that the most widely accepted theory of the Moon formation is the impact of the Earth by a Mars-sized body during the accretion period. Exceptionally though, volatiles are abundant in lunar soils. The lunar surface evolved during the heavy bombardment period, adding material with composition a different from that of the Sun. We have seen earlier that ions from the solar wind are directly implanted into the lunar surface [11]. This component was detected during the Apollo missions. The isotopic composition of the noble gases in lunar soils has been established as being subsequent to the formation of the Moon itself. In order to get further insights into the early Solar System, evidence has been searched for a predominantly non-solar origin of nitrogen in the convenient source of information that is represented by the lunar regolith[12]. This search suggests that, on average, some 90 percent of the N in the grains has *a non-*

solar source, contrary to the view that essentially all N in the lunar regolith has been trapped from the solar wind, but this explanation has difficulties accounting for both the abundance of nitrogen and a variation of the order of 30 percent in the $^{15}N/^{14}N$ ratio.

The Moon regolith presents a very challenging geological phenomenon. It consists of a very large number of grains with a rich history regarding their exposure to the Sun. Two parameters are useful in the systematic study of the lunar regolith: firstly, its "maturity" namely, the duration of solar wind exposure and, secondly the "antiquity", namely, how long ago the exposure took place. For the maturity parameter a useful way to measure it is in terms of the abundance of an element from the solar wind that is efficiently retained. The element nitrogen is a good example. (Alternatively solar noble-gas elements can be used.) Both antiquity and maturity have been used to learn about the evolution of the early solar system, especially the ancient Sun, the knowledge of which is needed for a comprehensive understanding of the problem of the origin of life on Earth. The exposure age to galactic cosmic rays produce certain nuclides in amounts proportional to the time the sample spends at the topmost part of the surface (some 2 m). The contract between the known low abundance of a certain nuclide and the one induced by cosmic rays produce an indicator of antiquity. The antiquity parameter has been discussed in detail[7]. Most of the N and some of the other volatile elements in lunar soils may actually have come from the Earth's atmosphere rather than the solar wind[13].

This hypothesis is valid provided the escape of atmospheric gases, and implantation into lunar soil grains, occurred at a time when the Earth had essentially no geomagnetic field. This is a valuable approach since it could clearly be tested by examination of lunar far side soils, which should lack the terrestrial component. This question is not just pertinent to the geological aspects of the evolution of the Moon, but by giving us a solid grasp on the evolution of the early Earth atmosphere, those factors that influenced the conditions favorable to the origin of life on Earth will be clearer.

7.8 Space climate and the early earth

During the early stages of the study of the origin of life[14,15] not enough attention was paid to the question of the correlation of chemical evolution on Earth and the all-important evolution of the still-to-be understood early Sun[16]. Today, due to the advent of a significant fleet of space missions and the possibility of performing experiments in the International Space Station (ISS), a meaningful study of the factors that may have led to an early onset of life on Earth begins to be possible.

We can understand general trends of the influence of space climate and weather on the evolution and distribution of life. Meteorites provide information on events that took place during this early period after the collapse of the solar nebula disk. Gas-rich meteorites have yielded evidence for a more active Sun.

A considerable number of young stars with remnants of accretion disks show energetic winds that emerge from the stars themselves. Similar ejections are still currently observed from our Sun. For this reason it is believed that some of the early Solar System material must keep the record of such emissions[17].

During the earliest stages of solar evolution, solar climate and weather presented an insurmountable barrier for the origin of life anywhere in the Solar System. In the Hadean, conditions may still have been somewhat favorable, especially with the broad set of ultraviolet defense mechanisms that are conceivable. The high ultraviolet (UV) flux of the early Sun would, in principle, cause destruction of prebiotic organic compounds due to the presence of an oxygen-free atmosphere without the present-day ozone layer. However, possible defense mechanisms against the UV radiation for the early forms of life have been proposed in the past[18,19].

7.9 Extra solar radiation during the evolution of life

Gamma ray bursts are powerful explosions that are known to originate in distant galaxies, and a large percentage likely arises from explosions of stars over 15 times more massive than our Sun.

A burst creates two oppositely directed beams of gamma rays that race off into space. The Swift mission, launched in November 2004, contributed to determine recent burst rates. Such data allows the evaluation of life's robustness during the Ordovician (namely, the time interval between 510 and 438 million years BP). During this geologic period there was a mass extinction of a large number of species (440-450 million years ago). This was the second most devastating extinction in Earth history. Present evidence has led to the conjecture that the extinction was triggered by a gamma ray burst[20]. There is no direct evidence that such a burst activated the ancient extinction. The conjecture is based on atmospheric modeling.

Gamma ray radiation from a relatively nearby star explosion, hitting the Earth for only ten seconds, could deplete up to half of the atmosphere's protective ozone layer. With the ozone layer damaged, ultraviolet radiation from the Sun could kill much life on land and near the surface of oceans, disrupting the food chain.

7.10 The search for life in the Solar System

We could gain deeper insights into the origin of life from the discovery of a second line of independent biological evolution within the Solar System. In such a case we could probe for its biosignatures with spacecrafts that are possible to construct with present-day technology. Indeed, some preliminary studies of such undertakings have already been completed under the auspices of ESA[21]. The Jovian satellite Europa continues to be a likely target for the search of biosignatures (cf., Sec. 12.40)[22].

Knowledge about the Jovian system grew from two sets of spacecrafts: the Pioneers 10 and 11 launched in the years 1974-1975 and Voyagers 1 and 2 that were launched in the summer of 1979. The Pioneer spacecraft probed the radiation environment of Jupiter and studied the main characteristics of the Jovian environment. When the Galileo mission entered Jupiter's orbit, further information on the radiation environment to which the satellites Io and Europa are exposed was inferred. Pioneer 10 flew by Jupiter in December 1973, the first space probe to do so, and discovered a large magnetic tail of Jupiter's magnetosphere.

What is more relevant from our point of view is the discovery of micron-sized dust particles [23]. The impact detectors of Pioneers 10 and 11 spacecrafts recorded 15 impacts of dust particles in the neighborhood of Jupiter, but at that time it was suggested that the source of some of the dust particles was due to comets [24]. The Ulysses spacecraft flew by Jupiter in 1992[25]. The Jovian system was recognized as a source of discontinuous streams of sub-micron dust particles. The Galileo Mission detected streams within 2 AU from Jupiter, as the mission began its measurements on arrival at the Jovian System[26]. Voyager had observed small dust in abundance at two places. Firstly, just as in the case of the well-known structure of Saturn, Jupiter was found to have a ring at a distance comparable with 1.8 times the radius of Jupiter (R_J)[27]. Secondly, Voyager also observed Io's volcanic plumes that were found to reach heights of about 300 km. Electromagnetic interaction of the particles making up the dust streams was evident on two separate occasions. The Ulysses and Galileo data revealed dust particles when these spacecraft were outside the Jovian magnetosphere. The arrival direction showed significant correlations with the ambient interplanetary magnetic field[25,26]. It was subsequently demonstrated that only particles in the 10^{-8} m size range can couple strongly enough to the interplanetary magnetic field to show the effects that were observed by Ulysses[28]. The corresponding impact speeds were deduced to be in excess of 200 km/s and the size of the particles was found to be in the range of 10 nm. Similar size dust particles were detected by the Galileo mission that originated from the inner Jovian system within several R_J from Jupiter[26,29]. These streams are strongly interacting with the planet's magnetic field[30].

The subsequent flyby of the Cassini-Huygens mission closest approach to Jupiter was at a distance of 137R_J, whereas at the same time (December 30, 2000), the Galileo spacecraft was at about 14 R_J. This occasion allowed simultaneous measurement of the Jovian dust streams. Dust trajectories exist that intersect both the Galileo and the Cassini-Huygens orbit. These measurements provided a direct measure to determine the time-of-flight of the grains between about 14 and 137 R_J. The above remarks illustrate the complexity of the interaction between the dust particles arising from Io and the Jovian magnetosphere[24]. This relationship will also be significant for the forthcoming missions to Europa (cf., Sec. 12.4), since the origin of the particles on its icy surface remain to be identified.

Further research is necessary to feel confident that we understand the distribution of the sulfur arising form Io in the neighborhood of Europa, as well as elsewhere in interplanetary space. Sulfur is the most likely chemical element that would show any evidence of biogenic signatures through the standard formulae of biogeochemistry (the study of biological effects by the consideration of isotopic fractionation of certain chemical elements)[31]. Thus any eventual space mission that would be capable of deciphering the isotopic fractionation of sulfur *in situ* on the icy surface of Europa could constrain considerably the differences between the various hypotheses regarding the nature of the endogenic source of sulfur.

In this chapter we have attempted to confront the reader with a flavor of the complex nature of the discussions that are relevant for a careful discussion of the search for a second Genesis. We need to understand how the conditions of the early Sun combine with observations in several disciplines to give us insights into the factors that lead to the habitability of the Solar System. Some of the areas of science that contribute to this research are biogeochemistry, lunar science, micropaleontology and chemical evolution.

Chapter 8

Towards the Intelligibility of Life

In Chapter 7 we extended the range of scientific disciplines that that may be invoked in our efforts to come to grips with the intelligibility of nature. Outstanding amongst these areas were disciplines that are familiar in astronomy. These included nucleosynthesis, biogeochemistry, space weather and solar physics. In this brief chapter we illustrate to what extent convergence is also an intrinsic feature in the life sciences.

The reader is advised to refer especially to the following entries in the Glossary: Contingency, Evolution (biological, cosmic and Lamarckian), Purines and Pyrimidines.

8.1 The universal driving force of evolution

The question of evolutionary convergence in the context of the life sciences has been discussed extensively[1-3]. What is more recent is our growing awareness that convergence is also a phenomenon that seems to be taking place in the space sciences, as explained in detail in Chapter 7 and elsewhere[4]. In the context of the life sciences assume that natural selection is the main driving force of evolution in the universe, a hypothesis made earlier elsewhere[5]. For these reasons it is relevant to question whether local environments that were favorable for the emergence of life on the early Earth, were at all unique, occurring exclusively in our own solar system.

Another view on universal validity of biology in the cosmos has been advanced in the context of its basic building blocks, the proteins made up of its monomeric amino acids and nucleic acids that are biopolymers of monomers that are called purines and pyrimidines[6]. It seems likely that the basic building blocks of life anywhere will be similar to our own. Themes that are suggested to be common to life elsewhere in the cosmos are the capture of adequate energy from physical and chemical processes to conduct the chemical transformations that are necessary for life: photosynthesis and chemosynthesis. Other factors that favor the universality of biochemistry are physical and genetic constraints. Beyond the universality of biochemistry in Chapter 10 we discuss the case in favor of assuming that the science of biology is universal, subject to experimental confirmation[7].

8.2 General aspects of convergence in biology

In general we may say that features that become more, rather than less similar through independent evolution, will be called "convergent". In fact, convergence in biology is often associated with similarity of function, as in the evolution of wings in birds and bats. New World cacti and the African spurge family provide an example.

Some further examples are the euphorbs, such as *Euphorbia stapfii*, and some members of the Madagascar Didieraceae *(Didiera madagascariensis)*. These plants are similar in appearance, being succulent, spiny, water storing, and adapted to desert conditions[8,9]. However, they are classified in separate and distinct families, sharing characteristics that have evolved independently in response to similar environmental challenges, and hence we may say that this is a typical case of convergence.

On the other hand, adaptive radiation is a second Darwinian concept that will be necessary for understanding the concept of restricted predictability in biology. In fact, adaptive radiation simply means evolution of an animal, or plant group into a wide variety of types adapted to specialized modes of life. In other words, adaptive radiation signifies evolutionary diversification of a single lineage into a variety of species with different adaptive properties.

Darwin's finches provide the classical example of adaptive radiation. Thirteen species of Darwin's finches live in the Galapagos Islands. They differ in the shape of their beaks. It is remarkable how versatile their beaks can be: keratin is the substance from that they are made. Evolutionary pressures can easily mould it. Thereby the origin of all the species now inhabiting these islands is facilitated. Besides, there is an additional species inhabiting Cocos Island in the Costa Rican territorial waters in the Pacific Ocean, north of the Galapagos Islands. We return later to some less familiar examples of adaptive radiation that will argue in favor of certain degree of predictability in biology.

8.3 Convergence at the biochemical level

The question *"What would be conserved if the tape of evolution were played twice?"* is relevant to astrobiology. It has been raised repeatedly in the past[10] and is used elsewhere in the present book. Since all forms of life known to us are terrestrial organisms, it is relevant to enquire whether the science of biology is of universal validity[7]. The sharp distinction between chance (contingency) and necessity (natural selection as the main driving force in evolution) is relevant for astrobiology. Independent of historical contingency, natural selection is powerful enough for organisms living in similar environments to be shaped to similar ends.

Our examples will favor the assumption that, to a certain extent and in certain conditions, natural selection may be stronger than chance[11]. We should look more carefully to the extent that convergence is built into biochemistry. Indeed, convergent evolution is ubiquitous. For instance, it manifests itself at the active sites of enzymes, in whole proteins, as well as in the genome itself. A few examples will help to illustrate the ubiquity of convergent evolution in the Earth biota:

- The northern sea cod (*Boreogadus saida*, Svalbard Norway) is an economically important marine fish of the family Gadidae. It is found on both sides of the North Atlantic. The distantly related order Perciformes with its suborder Percoidei, contains the sea basses, sunfishes, perches, and, more relevant to our interest, the notothenioid fishes from the Antarctic (*Dissotichus mawsoni*, McMurdo Sound, Antarctica). In spite of their distant relationship with cods, they have evolved the same type of antifreeze proteins, in which there are repeats of the same three amino acids[12]. These proteins are active in the fish's blood and avoid freezing by preventing the ice crystals from growing larger. The Antarctic fish protein arose over 7 Myr BP, while the Arctic cod first appeared about 3 Myr BP (both species arose in different episodes of genetic shuffling).
- The blind cavefish *Astyanax fasciatus* are sensitive to two long wavelength visual pigments. In humans the long wavelength green and red visual pigments diverged about 30 Myr BP. The mammalian lineage diverges form fishes about 400 Myr BP, but a recent episode in evolution has granted fish multiple wavelength-sensitive green and red pigments. Genetic analysis demonstrates that the red pigment in humans and fish evolved independently from the green pigment by a few identical amino acid substitutions[13], a clear case of evolutionary convergence at the molecular level.

8.4 The birth of chemical evolution

In Sec. 8.1 we raised the question of the possible universality of biochemistry, one of the sciences supporting chemical evolution. It seems likely that the building blocks of life anywhere will be similar to our own. Amino acids are formed readily from simple organic compounds. Stanley Miller first clearly and convincingly demonstrated this cornerstone of astrobiology. He was only a second year graduate student at the University of Chicago, when he published a remarkable paper in 1953 on the generation of amino acids. It was a simple experiment that attempted to reproduce conditions similar to those in the early Earth, when life first originated. He was under the guidance of Harold Urey, who had done fundamental research in nuclear physics. Urey was responsible for the discovery of an isotope of hydrogen: deuterium. He received the Nobel Prize for this work. Urey had subsequently suggested that the early Earth had conditions favorable for the formation of organic compounds. As a subject of his doctoral thesis Miller demonstrated experimentally that amino acids, the building blocks of the proteins could be formed without the intervention of man in environmental conditions, which may be called "prebiotic"—similar to those that presumably were reigning at the earliest stages in the evolution of the Earth itself.

8.5 The universality of the life sciences

Some themes are should be common to life elsewhere in the cosmos. For example, the capture of adequate energy from physical and chemical processes to conduct the chemical transformations that are necessary for life; in addition, photosynthesis and chemosynthesis could also be common features to life elsewhere. This expectation may guide the preparation for the search of bioindicators either in the Solar System or in other exo-planetary systems.

Other factors in favor of the universality of biochemistry are physical constraints (temperature, pressure and volume), as well as genetic constraints. Divergence and convergence are two evolutionary processes by which organisms become adapted to their environments. Evolutionary convergence has been defined as the acquisition of morphologically similar traits between distinctly unrelated organisms[13]. While there are many examples of molecular divergence, the same is not true of molecular convergence. Convergent evolution is said to occur when a particular trait evolves independently in two or more lineages from different ancestors[14]. This is distinct from parallelism, which refers to independent evolution of a trait from the same ancestor.

Although many of the best-known examples of convergence are morphological, convergence occurs at every level of biological organization. Another aspect of convergence that has been studied is "functional convergence". This phenomenon refers to molecules that serve the same function but have no sequence or structural similarity and carry out their function by entirely different mechanisms.

We cannot overemphasize the relevance of convergence and a further example will make this point clearer: structural convergence is another manifestation of the phenomenon referring to molecules with very different amino acid sequences that can assume similar structural motifs, which may carry out similar functions. In protein evolution, sequence divergence, rather than sequence convergence is the rule. In sequence convergence, one or more critical amino acids, or an amino acid sequence of two proteins, come to resemble each other due to natural selection. If the ancestral amino acids at a particular site were different in the ancestors of two proteins that now share an identical residue at that location, then convergent evolution may have occurred. (We shall return to the remarkable topic of convergence in more detail in Sec. 10.3.)

Throughout this chapter we have assumed that natural selection seems to be powerful enough to shape terrestrial organisms to similar ends, independent of historical contingency. This has an important implication to the main theme of this book. If we consider an extraterrestrial environment, it could be argued that the evolutionary steps that led to human beings would probably never repeat themselves, or that the probability of receiving a contact signal in the SETI project is very low; but those are hardly relevant points: the role of contingency in evolution has little bearing on the emergence of a particular biological property.

The phenomenon of evolutionary convergence has been illustrated with examples taken from biochemistry, although its occurrence extends to other branches of the life sciences. The assumed universality of biochemistry strongly militates in favor of selecting biomarkers in Solar System exploration from standard biochemistry. To drive this crucial point home, in the next section we shall provide additional examples of convergence in the life science, this time taken from animals such as mollusks, birds and fishes.

8.6 Convergence at the level of animals

The evolutionary biology of the Bivalvia, both at the level of zoology and paleontology, provide multiple examples of convergence and parallel evolution, a fact that makes difficult the interpretation of their evolutionary history[15-16]. We should recall that amongst the vertebrates, Passeriformes is a group of birds (including swallows) that may be confused with Apodiformes (including swifts), but that they are not related to them.

Swallows and swifts provide a classical example of evolutionary convergence. Members of these two orders differ widely in anatomy and their similarities are the result of convergent evolution on different stocks that have become adapted to the same life styles in similar ecosystems for both species[2].

Another example, this time from an aquatic environment has been pointed out involving two groups: tunas and lamnid sharks. The second group belongs to the family Lamnidae. These sharks inhabit tropical to cold temperate waters in almost all seas.

They are predators and have similar streamlined body shapes. The two groups have developed high-speed swimming techniques that involve efficient movements of only their tail sections, rather than the full-body undulations of most fish[17]. While the two groups' common ancestor diverged more than 400 million years ago, selection pressures seem to have made both fish hit upon nearly identical movement mechanisms.

More striking still is the fact that an anatomical feature reveals evolutionary convergence with tunas, namely the location of a specific muscle (the red, aerobic, locomotor muscle, RM). It is deep in the body next to the vertebral column and concentrated in the mid-body region, in contrast with the lateral and subcutaneous position of RM in other fishes[18].

Coupled with vascular heat exchangers, tunas and lamnid sharks have evolved the ability to retain metabolic heat in RM and attain temperatures that are significantly above that of the surrounding water.

These examples further illustrate convincingly that convergence in the life sciences is a well-established phenomenon.

Chapter 9

Is Life Ubiquitous in the Universe?

Having been encouraged so far by the insights that evolutionary convergence provides in our attempts to understand the intelligibility of nature, we now enquire whether life is widespread in the universe. Does the scientific method together with appropriate technologies allow us to decipher essential aspects of the intelligibility of nature? In this chapter we concentrate on the discussion of a conjectured universal evolutionary pathway to cellular evolution, multicellularity, the origin of neurons and their connections, cerebral ganglions and eventually to brains and intelligence. This point of view advocates the use of standard optical and radio techniques for the search of extraterrestrial life. But in our cosmic backyard the exploration of the Solar System may yet help us to decipher aspects of the intelligibility of nature.

The reader is advised to refer especially to the following entries in the Glossary: Augustine of Hippo, Cassini-Huygens, Copernicus, Eukaryogenesis, Giordano Bruno, Hydrothermal vent.

9.1 Introduction

Comparative analyses enable some questions regarding limited aspects of brain evolution to be answered[1]. Differences in the neocortex consistently take the form of module formation. For this reason changes in, for instance, internal organization as well as changes in connection patterns suggest that critical alterations in brain evolution are generated by similar mechanisms. Such mechanisms were probably present very early in mammalian evolution. Since mammals represent a direct descent of primitive multicellular animals, these remarks have to be evaluated seriously, due to their relevance in the appearance of intelligent life elsewhere.

Indeed, it is conceivable that the common mechanisms are so fundamental that future changes in human brain evolution will be governed by similar mechanisms, as has been conjectured already by evolutionary neuroscientists. Similarly, within the context of astrobiology, we take one step further and conjecture that evolution elsewhere in the universe is governed by the same (universal) mechanisms characteristic of brain evolution on Earth[2].

However, it is important to rationalize the exceedingly variable cognitive abilities observed across species, for they can be understood, in spite of the existence of the conjectured mechanisms that are underlying brain evolution; for instance, small contingent changes that occur in the genome may be the source of the diversity in the evolution of the neocortex in mammals. Our conjectured evolutionary pathway offers the basis for introducing some rationalization in the planning of experiments that are worthwhile, and feasible, in the forthcoming campaign of missions to Europa and Mars. Within this context, in the present work we review one class of possible experiments.

We restrict the consequences of the conjecture of the universal pathway to intelligence to two aspects. Firstly, its possible implications regarding biological evolution outside habitable zones in solar systems and, the likelihood of the onset of other intelligence. Secondly, the present discussion gives us an opportunity for rationalizing the dialogue that should be maintained with philosophers and theologians with respect to the incorporation into our cultural background of a new view on our position in the universe. By now, most of us agree on what is the subject matter of astrobiology, which includes[3]:

1. origin,
2. evolution,
3. distribution and
4. destiny of life in the universe.

These are well-established research areas, as we have tried to illustrate in earlier chapters. Darwin had made a contribution with respect to the evolution of life, rather than its origin. We will pay special attention to the last two topics of astrobiology: distribution and destiny of life in the universe.

Our next task is to consider the problem of the distribution of life in terms of the physical, earth and life sciences. Philosophers from the time of Plato have also thought about these issues. Although our activities in science and the humanities are intrinsically different, a dialogue is of mutual benefit. Returning to the distribution problem, we may ask:

9.2 Is intelligence a universal phenomenon?

First of all, the only intelligent life that we are aware of in the universe is life on our own planet. We have no evidence of how representative it might be of life elsewhere, at least suitable environmental conditions, has been discussed in the literature[4].

Two of the basic principles of terrestrial biology are natural selection and the existence of a common ancestor for all life on Earth. This can be interpreted as evidence in favor of the fact that at a cosmic level evolution is predictable to a certain extent, and it is not entirely dependent on chance. The underlying question concerns the relative roles of adaptation, chance and history, a topic that is, strictly speaking, subject to experimental tests. As seeing in the previous chapter further support from independent teams suggests that natural selection overrides the randomness of genetic drift. In other words, natural selection seems to be powerful enough to shape terrestrial organisms to similar ends, independent of contingency. The relative importance of chance and necessity continues to ignite heated debates in the scientific community.

On the other hand, some arguments suggest that a human-level of intelligence is a consequence of a conjectured universal constrained evolutionary pathway towards intelligence as it occurred on Earth. Certainly, in an extraterrestrial environment the evolutionary steps that led to human beings would probably never repeat themselves; but as pointed out earlier in the book that is hardly the relevant point: the role of chance in evolution has little bearing on the emergence of a particular biological property.

As we have seen in Chapter 8, the inevitability of the emergence of particular biological properties is a phenomenon that can be linked to evolutionary convergence that was defined in Sec. 8.4. In order to demonstrate the ubiquity of this singular phenomenon o the evolution of life on Earth, we add an additional example, but now we take it from the neurosciences:

• Brain structures may look the same, but in some cases they may have arisen independently, in different lineages. In such cases we may say that have arisen by convergent evolution. In particular, neurons can be classified as to whether they are driven by one eye or the other. This property is called "ocular dominance". From a sketch of the organization of the visual pathway, we can appreciate that when an electrode penetrates the cortex, all neurons encountered in the column defined by the penetration, have the same property of being driven by one eye or the other. This has been observed in cats and monkeys. Hence, this property is likely to have arisen form convergent evolution, since cats and monkeys are very distantly related species, and other intervening lineages do not posses this feature.

In the light of this the conclusion that our intelligence is unrepeatable is no longer so evident: The presumed uniqueness of the phenomenon of intelligent behavior depends on a combination biological, geological and possibly other factors. In simpler terms, the metaphor that was cited in Sec. 8.3:

> *"if the history of evolution were to be repeated, it would not lead to man again."*

has evidently to be taken with a degree of caution and humility. The relevant question to be pondered upon in the context of astrobiology, as well as philosophy and theology is whether intelligence may emerge elsewhere. The emergence of humans is not necessarily the only topic that can encourage the interdisciplinary dialogue between science and the humanities.

In order to ask the pertinent question, we must appeal to the space sciences through, for instance, the SETI project. These considerations raise additional questions. For instance, what can present-day neuroscience tell us about the pathway towards intelligence from what may be observe in the evolution from bacteria to humans?

The coelenterates, such as the medusa, form the first taxon, or group, in the evolution of multicellular animals. In these organisms neurons and true nervous systems appear for the first time in terrestrial evolution. There are individual interconnected neurons transmitting, chemically, or electrically, in both directions, forming a sort of nerve net[5]. Hence, for life on Earth, neurons appear at almost the lowest level of multicellularity. These arguments strongly advocate in favor of the existence of other human-level of intelligence, elsewhere in the cosmos. Essentially, the argument hinges on the transition from prokaryotes to eukaryotes a question that may be tested in the future if the current programs of the main space agencies identify a second Genesis.

9.3 Is cellular evolution bound to take place?

The main reason why we believe that, for instance, eukaryogenesis may be universal[6,7] is that not only physico-chemical processes regulate the structure, and evolution of the universe; but besides, we may assume that all principles for understanding nature are of general validity, irrespective of the location of the observer in a given galaxy. In particular, this "strong" principle of universality of the natural sciences, forces upon us the consideration that the scientific bases of biology, as known to us on Earth, also apply elsewhere. This question was discussed in the previous chapter.

According to such universality of the natural laws, evolution of biological complexity in our cosmic environment is constrained by the two basic principles of biology: natural selection and common descent. But there are compelling reasons to raise the question of biology in our Solar System, especially in the Jupiter system. Amongst other factors, there is some evidence in favor of the habitability of this Galilean satellite. One of the motivations for this hope will be taken up again in Chapter 12, where we will comment the likely presence of an ice-covered ocean. This possibility has been gradually forced upon us by measurements from the Galileo mission (cf., Sec. 7.10). In fact, the possibility Europa's habitability is independent of the likely existence of hydrothermal vents at the bottom of the ocean.

It is conceivable that independent of whether there are hydrothermal sources at the bottom of the ocean, a thriving community of microorganisms might have evolved in that environment anyway. This possibility does not require the technological difficulties of research from a submersible that would search out hydrothermal vents. A lander would be sufficient to process melt water that may already contain microorganisms due to resurfacing of the satellite (the renewing of the satellite surface). This property is evident from the image on the cover of this book, where it is evident that there are not too many craters, as in the case of our own Moon.

9.4 Where can we search for a second Genesis?

There are several cases of some interest:

- Europa may be a typical biofriendly body amongst extra-solar planets. The main problem is how to select appropriate experiments in situ, after surface landers are able to filter melt water from Europa's frozen surface.

- Evolutionary tests are also suitable in potential water sources under the Martian surface, such as in its north pole. These environments may contain microorganisms, an exciting possibility that was raised by the Allan Hill meteorite. Although the possibility of having detected life in this meteorite was not subsequently confirmed, it was nevertheless an important contribution that stimulated discussion about the possibility of life elsewhere in our Solar System. Cases of possible habitability to test in the future have been discussed in the literature[5].

- Titan is a potential cradle for life. After the early successes of the Cassini-Huygens Mission, many interesting exobiological questions have been raised, including the source of methane, and possible ammonia-water ocean[8], inside this large satellite that resembles the Archaean Earth in some respects. For instance, Titan has a nitrogen atmosphere, and so does the Earth, including its atmosphere before life. Titan has organics that are almost certainly supplied in the absence of life. Not all the Earth's prebiotic ingredients are present on Titan though, because the Earth probably had CO_2 unlike Titan.

• Cassini has confirmed that icy jets shooting up to 500 km are ejected from Enceladus, a tiny satellite of Saturn. The presence of liquid water in its interior raises this moon to a prime candidate for the search for life. Other reasons for focusing on Enceladus is that Cassini flew within 175 km in 2005 confirming the presence of an atmosphere: their instruments found that the atmosphere contains water vapor comprising up to about 65 percent, with molecular hydrogen at about 20 percent. The rest is mostly carbon dioxide and some combination of molecular nitrogen and carbon monoxide. Another Cassini instrument showed that the south pole is warmer than near the equator. The poles should be colder because the Sun shines so obliquely there. However, in small areas of the pole, concentrated near the fractures known as the "tiger stripes", the temperatures can reach well over 110 Kelvin (-261 F). This should be compared with the equatorial temperature of 80 degrees Kelvin.

From the above four examples we have learnt that searching for a second Genesis is feasible in the foreseeable future. The ultimate aim of a biology experiment in Solar System exploration is to develop robotic tests that are compatible with the necessarily reduced dimension of landers. In the case of a Martian mission to subterranean pockets of liquid water we find a possibility that has attracted wide attention both by scientists and by the popular press. For this purpose miniaturization of instrumentation is essential. Many difficulties though are inherent in the eventual design of a test that would intend to identify microbes, robotically, in any extraterrestrial environment. This question begins to be important, in view of the decisions that have to be made in the selection of biological experiments that should be performed in situ on Europa and on Mars.

9.5 Habitability in the Solar System

The arrival of the Galileo mission in the Jovian system gave rise to a NASA meeting at San Juan Capistrano and related publications, where the habitability of Europa was discussed, both in terms of microbiology[9], as well as in terms of novel instrumentation for missions to the Jovian system. Early discussions, long before the proposed missions for the future (to be discussed in Sec. 12.4), also considered the possibility exploring Europa's habitability in the future in terms of direct use of a submersible called a hydrobot[10].

This question is still relevant a decade later, in terms of new NASA tests of an autonomous underwater vehicle (AUV) called ENDURANCE for the Astrobiology Science and Technology for Exploring Planets (ASTEP) program[11], a worthy successor of our cryobot-hydrobot early planning. Significant papers have much more recently continued the conditions for establishing a stable ecosystem. They include discussions of the biochemistry[12,13], as well as the biogeochemistry of the habitability of Europa[14,15].

In relation with the distribution and destiny of life in the universe we have argued that if the experiments on evolution were to be successful, the science of the distribution of life in the universe would lie on solid scientific bases. Given such bases for the distribution of life in the universe, it does not seem premature to include in our discussions other sectors of our society.

9.6 Implications of life in the cosmos in the humanities

Science and religion are both concerned with the common understanding of the destiny of life in the universe. These two intrinsically different cultural activities largely address the same questions. For this we should all persevere in establishing a constructive dialogue.

Giordano Bruno was a pioneer in this inter-cultural dialogue[16]. In 1584 he wrote his first three Italian dialogues during a visit to England[8]. These writings were stimulated by debates at the University of Oxford. In his third dialogue: *"On the infinite universe and worlds"*, Bruno introduced concepts that are still at the centre of astrobiology. He suggested the removal of the Copernican "sphere of fixed stars".

Bruno spoke of an infinite universe and the plurality of inhabited worlds. Such a cosmological vision matured in Bruno's writings long before the science of astrometry allowed this concept to be brought within the scientific domain.

The large number of new extra-solar planets underlines Bruno's remarkable achievement. The position of Augustine was shared by Bruno and soon after it was also shared by Galileo. This view was demonstrated in Galileo's writings, for instance, the *"Starry Messenger"*. The majority of contemporary thinkers now share the Augustinian viewpoint. This progress may open the way to further fruitful dialogue on the question of the destiny of life in the universe. In spite of all the progress in astrobiology during the last century, some of the answers to key questions of common interest are escaping us. This is a concept that is of interest to all forms of culture.

From the *"Starry Messenger"* we may take, as an illustration, Galileo's sketch of the constellation of Orion. His simplicity is matched by our still-too-simple view of life on Earth, which is summarized in the three domains of the phylogenetic tree of life on Earth.

Elsewhere in this book we have argued that at the microscopic level our tree of life may be of universal validity.

9.7 Ultimate implications of evolution in the cosmos

Such experiments would undoubtedly have a significant impact in our culture, not just in our scientific outlook. The influence of the new knowledge will also be felt while discussing the deep questions raised in the past by George Coyne[9] and Robert Russell[17], such as the concepts of creation, morality and redemption. These questions may be discussed not just against a background of our particular evolutionary line, which has been followed up by life on Earth. Such issues ought to be discussed already in terms of the many evolutionary lines that are hinted at by astrobiology.

Chapter 10

Testing the Universality of Biology

Our aim in previous chapters has been to consider the likelihood and implications of a second Genesis. What about testing this conjecture? This question is of paramount relevance for astrobiology. The main point of this chapter is to discuss evolution itself as the result of two competing factors: contingency and convergence. A substantial body of evidence argues in favor of evolutionary convergence having played a major role in the Earth biota during its ascent from bacteria to humans. Following the seminal contribution of Darwin, it is reasonable to assume that all forms of life known to us so far are not only terrestrial, but are descendants of a common ancestor that evolved on this planet at the end of a process of chemical evolution. We also raise the related question of whether the molecular events that were precursors to the origin of life on Earth are bound to occur elsewhere in the universe.

The reader is advised to refer especially to the following entries in the Glossary: Biogeochemistry, Biomarker (biosignature), Biota, Cassini-Huygens mission, Convergent evolution, Galileo mission, Stellar evolution and Voyagers. He is also recommended to read again the entry for "Evolution (biological)".

10.1 Contingency and convergence in evolution

Beyond the specialists of the theory of evolution the question of the relative importance of chance and necessity was brought to the attention of a large number of scientists by a well-known book[1]. The main issue is which features of the history of life are inevitable and which are highly contingent and, therefore, unpredictable. There is a broad list of publications addressing this issue. Following the publication of a series of books[2-4], especially interesting discussions have been published.

These books, and the ones cited above, have discussed extensively the question of the relative importance of contingency and convergence[5-9]. We begin the next section discussing some evidence that in spite of the intrinsic contingency of Darwinism, sometimes history tends to repeat itself during the course of evolution. Later, we explore the consequences of the hypothesis that such repetitions have in the search for life in the universe.

10.2 Evolutionary history tends to repeat itself

As we have illustrated in Chapter 8, the phenomenon of convergence occurs at various levels such as morphology, physiology, behavior, and even at the molecular level. Specifically, deep insights can be drawn from the neurosciences. The comparative approach of modern evolutionary neurobiology has been discussed in detail[10]. Neuroscientists often work with animals instead of humans for discovering the principles of neural organization ("the comparative approach"). Homology refers to a structure, behavior, or even a gene that has been retained from a common ancestor.

A clear example of homology is a wing of a bat, as well as the hand of a human; both of them have a common evolutionary descent. What is more interesting from our point of view is that the wing of a bat and the wing of an insect are not homologous, but they are examples of evolutionary convergence. (These structures look the same but have not a common descent.) One example of convergent evolution in the brain was explained in Sec. 9.2. What we learn from them is that they demonstrate the limited and rigid rules by which brains evolve. Indeed, we can go further recognizing that evolutionary convergence allows the examination of limitations inherent in constructing nervous systems.

For these reasons, evolutionary convergence is, consequently, a significant phenomenon that should be discussed in the context of the universality of biology (namely, for demonstrating that biology is a science that is not only confined to the Earth biota, and hence it is a science of universal validity). The topic of the repetition of evolutionary history will be illustrated below with two examples taken firstly from the evolution of the brain, and secondly from biochemistry.

We begin, as mentioned earlier, with an example on brain in the dolphin[11,12]. Although evolution has been associated with a "tinkerer", detailed considerations of the products that evolution constructs point out that there are a limited number of underlying mechanisms that are accessed for building brains. The unit in question is a "module", that is a structure that not only occurs in large-brained mammals, such as dolphin and humans, but it also similar to that which occurs in small-brained mammals, such as the mouse. This suggests that module-size is preserved by evolution across species. The implication of this remark is that in the course of evolutionary history repetition, in fact, does occur. In other words, evolutionary history tends to repeat itself.

We can go further: these examples (and many others that we have not included for lack of space), suggest that the human brain is enslaved within the same genetic constraints as the brains of other mammals. Consequently, its future evolution will be expected to follow constraints imposed on evolving nervous systems that are gradually being discovered by comparative neuroscience studies that have exposed similarities in cortical organization across species[13]. The mechanisms for possible changes are guided by the same mechanisms that were responsible for the overall structure of other mammals.

To sum up, many structures, some of which had previously been assumed to be homologous, have evolved many times independently. These studies demonstrate that evolutionary convergence in brain anatomy and function is widespread.

There is an analogous illustration for the widespread occurrence of evolutionary convergence in biochemistry: Darwinian evolution has been shown to follow only very few mutational paths to fitter proteins. In this case, once again contingency is limited by a diverse variety of constraints that are imposed on the evolutionary process.

Such selective inaccessibility implies that the replaying of the tape of life at the biochemical level might make protein evolution not only repetitive, but even predictable[14].

As we are assuming that evolutionary convergence is widespread phenomenon in (universal) biology we do not dwell on the question of whether other life uses some other structure. Indeed, this work provides another example that the number of mechanisms that are accessed by natural selection is, in a number of cases, a limited set. We should compare this result from biochemistry with the above illustration that large brains are constructed in a similar fashion, independent of recent evolutionary history.

Multiple instances of "history repeating itself" abound in the life sciences, for instance the science of paleontology highlights the morphological analogies of organisms that live in similar environments in order to interpret their corresponding paleo-environments[15].

The morphological similarities are evident in sessile benthos between the coral *Omphyma* and the bivalve *Hippurites*. Paleontology also provides us with other remarkable examples of evolutionary convergence, as illustrated by the characteristic body shape of swimming vertebrates (the shark *Lamna* and the mammal *Focaena,* both of which are known morphologically as "torpedo-type").

We have discussed organisms and anatomic structures that have been discovered in the construction of a biological system by the action of natural selection. Recently we, and other authors, have attempted to explore the implications of evolutionary convergence for the consideration of the emergence of biology anywhere in the universe[16,17]. If the ubiquity of evolutionary convergence is correct, from the point of view of Darwinism contingency and evolutionary convergence are opposite competing factors that have to be evaluated together.

In Chapter 12 we discuss experiments that intend to identify biological indicators elsewhere in our own Solar System. We shall pay special attention to one of the leading candidates for habitability, the Galilean moon Europa.

10.3 The ubiquity of convergent evolution

Evolutionary convergence strongly advocates in favor future space missions aiming to probe the universality of biology by careful tests based on the biochemistry that we already know. As we shall see in Chapter 12 at the present time the number of space agencies is increasing. The major agencies are in the United States, Europe, Russia, Japan, China and India. Most of their efforts are in frontier technology. A minor, but significant part of their work is at the search for a second Genesis. The question of convergent evolution is significant, as it may guide our search for biomarkers with the biology that we do know on Earth.

Divergence and convergence are two evolutionary processes by which organisms become adapted to their environments. Evolutionary convergence has been defined as the acquisition of morphologically similar traits between distinctly unrelated organisms[18]. Although many of the best-known examples of convergence are morphological, as mentioned above convergence occurs at every level of biological organization. However, molecular convergent evolution is most relevant for our enquiry whether life is a universal phenomenon, and from the point of view of this review, we should also consider biochemical convergence in some detail (cf., the related discussion on convergence in "The universality of the life sciences" in Sec. 8.5).

Certain types of enzymes (e.g., serine proteases) have evolved independently in bacteria (e.g., subtilisin) and vertebrates (e.g., trypsin). Despite their very different sequences and three-dimensional structures, they are such that the same set of three amino acids forms the active site[19]. Another example is structural convergence. This refers to molecules with very different amino acid sequences that can assume similar structural motifs, which may carry out similar functions. An example is the remarkable similarity observed in biomolecules that occur in the field of immunology[20].

Finally, one or more critical amino acids, or an amino acid sequence of two proteins may come to resemble each other due to natural selection. If the putative ancestral amino acids at a particular site were different in the ancestors of two proteins that now share an identical residue at that location, then convergent evolution may have occurred. The most frequently cited case of convergence and parallelism at the sequence level is a digestive enzyme (lysozyme) in a number of unrelated animals. This group includes the langur (a primate), the cow (an artiodactyls), and the hoatzin (a bird). All of these animals have independently evolved the ability to use bacteria in order to digest cellulose[21,22]. Here, a few specific residues have evolved in convergence to allow digestion of cellulose-eating bacteria.

10.4 The Solar System is not unique

In Chapter 7 we have attempted to document convergence in the space sciences. This phenomenon should include a series of well-established observations that point in the direction that our solar system and its galactic neighborhood are not unique in many respects that are particularly relevant for the emergence of life. In the present section we review a few of them to make our point sufficiently supported by numerous observations. Of paramount importance in this respect is the broad knowledge that we have gathered in the old subject of chemical evolution that has been well reviewed over the last decade as already mentioned.

The concept of cosmic convergence has a second aspect that may be inferred from what we know about small bodies, such as the Murchison meteorite. These bodies may even play a role in the origin of life. According to chemical analyses in this particular meteorite, which contains basic molecules that are needed for the origin of life, such as lipids, nucleotides, and more than 70 amino acids[23]. Most of the amino acids are not relevant to life on earth and may be unique to meteorites. This demonstrates that those amino acids present in the Murchison meteorite, which also play the role of protein monomers, are indeed of extraterrestrial origin. If the presence of biomolecules on the early earth is due in part to the bombardment of interplanetary dust particles, comets, and meteorites, then the same phenomenon could be taking place in other solar system.

The interstellar medium provides yet another illustration of convergent phenomena that occur at a cosmic level. Indeed, solar systems, many of which are now known, originate from interstellar dust that is constituted mainly of the biogenicl elements for life, such as C, N, O, S, P, and a few others. Stars as they evolve and go through the main sequence of the Hertzsprung-Russell diagram (cf., Sec. 6.1) expel their material into interstellar space in two different. Firstly, when stars of at least 0.4 solar masses exhaust their supply of hydrogen, their outer layers expand to form a red giant. Eventually the core is compressed enough to start helium fusion, gradually shrinking the star radius and increasing its surface temperature. After the star has consumed the helium at the core, fusion continues in a shell around a hot core of carbon and oxygen. After a series of intermediate steps the final stage is reached when the star begins producing iron. In relatively old and massive stars, a large core of iron accumulates in the centre of the star. An average-size star will then shed its outer layers as a planetary nebula. If what remains after the outer atmosphere has been shed is less than 1.4 solar masses, it shrinks to a white dwarf somewhat similar to the size of the Earth.

Secondly, in larger stars, fusion continues until the iron core has to more than 1.4 solar masses the core will suddenly collapse. The shockwave formed by this sudden collapse causes the rest of the star to explode in a supernova.

These supernovae explosions are a source of enrichment of the chemical composition of the interstellar medium. In turn, these events provide new raw material for subsequent generations of star formation, which leads to the formation of planets. Late in their evolution, stars are still poor in some of the heavier biogenic elements (for instance, magnesium and phosphorus). Such elements are the product of nucleosynthesis triggered in the extreme physical conditions that occur in the supernova event itself. By this means, the newly synthesized elements are disseminated into interstellar space, becoming dust particles after a few generations of star births and deaths[24]. In both cases of the evolution of stars, heavy elements may be recycled to form new stars and terrestrial (planets). In this sense there is universal convergence of terrestrial planet formation independent of a given solar system.

An additional case that argues in favor of convergence at a cosmic level is emerging from what we are beginning to learn about the origin of planetary systems around stars. Our solar system formed in the midst of a dense interstellar cloud of dust and gas, essentially a circumstellar disk around the early sun. Some evidence suggests that this event was triggered by the shock wave of a nearby supernova explosion more than 5 billion years ago. Indeed, additional evidence mentioned in Sec. 7.4 indicates the presence of silicon carbide (carborundum, SiC) grains in the Murchison meteorite, a fact demonstrating that they are matter from a type II supernova[25]. We may now be observing an extra-solar circumstellar disk around a young 3-million-year-old sun-like star in the constellation Monoceros[26]. Several earlier examples of circumstellar disks are known, including a significantly narrow one around an eight-million-year-old star.

The narrowness of this disk suggests the presence of planets constraining the disk[27]. The following additional information further supports the arguments in favor of universal mechanisms of convergence in the formation of solar systems; that is, the matter of the original collapsing interstellar cloud does not coalesce into the star itself, but collapses into the spinning circumstellar disk, where planets are thought to be formed by a process of accretion. Some planetesimals collide and stay together because of the gravitational force. In addition, a variety of small bodies are formed in the disk, prominent among which are comets, asteroids, and meteorites, completing the components that make up a solar system, as we know it.

Finally, the fifth example of cosmic convergence is provided by the convergent origin of hydrospheres and atmospheres. The earliest preserved geologic period (the lower Archaean) may be considered as representing the tail end of the heavy bombardment period (cf., Sec. 6.4). During that time, various small bodies, including comets, collided frequently with the early precursors of the biomolecules that eventually ignited the evolutionary process on earth and in its oceans. In addition, comets may be the source of other volatile substances significant to the biosphere, as well as the biochemical elements that were precursors of the biomolecules[28].

Collisions with comets, therefore, are thought to have played a significant role in the formation of the hydrosphere and atmosphere of habitable planets, including the earth. The source of comets is the Oort cloud and Kuiper belt. These two components of the outer Solar System seem to be common in other solar systems. Hence, in the sense of the above-mentioned examples, we recognize evolutionary convergence in a cosmic scale.

10.5 Can the universality of biology be tested?

Testing the nature of biology within the Solar System is gradually becoming more feasible with available technology. This is especially true, due to the new technology that is currently being developed. The search for life elsewhere in the universe is a time-honored research that was called bioastronomy in Sec. 2.13. A large number of researchers have followed up this discipline since the middle of last century. The SETI project has advanced at a vertiginous pace[29]. With the tools that will be available in the near future more definite searches specifically focusing on likely exoplanets.

For instance, the Convection Rotation and Planetary Transits (COROT) mission that has been supported by CNES, ESA, Austria, Spain, Germany, Belgium and Brazil, will search for rocky Earth-like planets. Later on the ESA DARWIN mission will be aimed at the search for planets and possible biomarkers on them in the mid-infrared. It will be possible with the DARWIN mission to study of nearby terrestrial exoplanets (< 25 pc) that will be orbiting stars within their habitable zone. Darwin is expected to launch in the 2015 timeframe.

In view of this technological promise, the central problem of astrobiology (the existence of life elsewhere in the universe) is no longer the exclusive domain of organic chemistry. As we have seen in Chapter 3, the field of "chemical evolution" was developed throughout last century[30]. This field has been extensively reviewed over the last decade[31-37]. We expect that radio astronomy and space exploration will be an ever-increasing stronger partner with a significant relevant role to play.

There are significant strategies for identifying those places where future landers could search for the biomarkers. The Galileo Near-Infrared Mapping Spectrometer (NIMS) provided some evidence for the presence of sulfur compounds has been discussed in elsewhere[38]. The most likely sites would be where the salt deposits, or organics, are concentrated, as suggested by the NIMS data. For instance, the search for biomarkers on Europa could focus on the area north of the equatorial region, between 0 and 30 N and between the longitudes 240 and 270[39]. But a more intriguing and smaller patch would be the narrow band with high-concentration of non-ice elements that lies east of the Conamara Chaos, between the Belus and Asterius lineae, namely, between 18-20 N, and longitudes 198-202. Definite answers can be searched in situ on the icy surface.

A specific example is provided by mass spectrometry on a possible future lander on Europa. At this stage it is possible to suggest the best possible landing site. We have suggested that at the "patch" found in the Europan surface coordinates 200W, 20N (longitude and latitude, respectively, there is a scientific valid way of testing biogenicity through isotopic fractionation that may have occurred on sulfur patches on the Europan icy surface[40-41].

As a guiding line in our search for a way out of the impasse created by still not having had a first contact with an extraterrestrial civilization, we assume, as a working hypothesis, that evolution of life in the universe can be explained only in terms of evolutionary forces that we experience today in our local environment: Although there are still many questions to be answered at present it seems possible (although not an easy matter) to penetrate the oceans of the iced Galilean satellites (cf., Sec. 9.5). Everyone agrees that the Newton's theory of gravitation can be extrapolated without any difficulty throughout the universe, except for the corrections implied in the theory of general relativity. The case of extrapolating the theory of biological evolution throughout the cosmos requires more care and is evidently still an open problem.

Arguments against the hypothesis of "biogeocentricism" (the view that maintains that life is confined to planet Earth) can now be formulated thanks to progress in our understanding of Darwinian evolution[42]. The role of randomness has been qualified since Darwin's time. The role of chance is implicit in *The Origin of Species*. We have also seen that molecular biology constrains chance. Evolutionary convergence is an additional factor to take into account, as illustrated with the above examples of neuroscience and biochemistry. To sum up, Darwinian contingency is constrained. Evolution often tends to converge on similar solutions when natural selection acts on similar organic materials that are in similar environments.

On the other hand, astrochemistry and planetary science present us a picture in which the environments where life can originate are limited and are supplied with analogous abundances of the chemical elements. We already are gathering information on a significant number of Jupiter-like planets around stars in our cosmic neighborhood. Such planets arise form sub-nebulae that are likely to yield an array of satellites around them. In our outer solar system this can be confirmed. Each of the giant planets in our solar system has a large suite of satellites. Factors giving rise to atmospheres in satellites of the giant planets, such as Saturn are known. Titan, for instance, has an atmosphere that was produced by out-gassing, combined by seeding of volatiles by comets carrying a fraction of water ice.

Evidence is leaning in favor of the existence of Jovian planets in our galaxy with masses larger than Jupiter. Hence, tidal heating responsible for Io's volcanic eruptions could be even more efficient in other solar systems. On Europa it is not completely clear that tidal heating may produce hydrothermal vents capable of giving rise to life, but the case in favor of this hypothesis is strong [43]. Tidal heating may be even more efficient on satellites orbiting around Jupiter-like planets with masses larger than the Jovian mass. Natural selection will be working in those extra-solar cases on a finite number of similar environments.

Once again, in astrochemistry, similar chemical elements will be available for chemical evolution. We have also learnt that there are no laws in chemical evolution that are specific to the Earth; it is reasonable to hypothesize that biological evolution will follow, once the molecules of life have emerged from chemical evolution. Darwinism cannot be seen simply as a dichotomy between chance and necessity, but constrained chance and convergent evolution will favor analogous pathways that have led to the evolution of life on Earth.

10.6 Searching for biosignatures in the Solar System

In this chapter we have not attempted to be exhaustive on the question of convergence, either biological or cosmic. Indeed, it is not even necessary, as there are excellent texts and papers that have already achieved this purpose. (Excellent books are already available, and were mentioned above.) The examples cited above are only meant to frame the question that the search for life makes sense with the biology and physics intuition that we have learnt on Earth (convergence in biology, and physics).

The main point of this chapter has been to discuss that evolution itself is the result of two competing factors: contingency and convergence. A substantial body of evidence argues in favor of evolutionary convergence having played a major role in the Earth biota during its ascent from bacteria to humans. The concept of convergence in the space sciences, discussed in Chapter 7, argues in favor that biology (including astrobiology) is a universal science.

The examples that we have cited highlight the ubiquity of evolutionary convergence. What does it mean within the framework of Darwinism? Indeed, the ubiquity of evolutionary convergence argues against biological diversity being unique to Earth and that within certain limits the outcome of evolutionary processes might be predictable.

We discussed in the previous chapter that the human brain seems to have the same genetic constraints as the brains of other mammals. Consequently, our expectation for its future evolution is that the brain will follow predictable paths that are guided by the same mechanisms responsible for the overall structure of other mammals. But clearly, the precise specializations that may emerge cannot be known[13].

These Darwinian arguments contribute to provide a cornerstone for our thinking on searching for evolutionary biosignatures during the exploration of the Solar System. Finding traces of life in any of the candidate-sites that are known to space geophysicists, such as Europa, Enceladus, Titan, or Mars would add arguments towards obtaining further insights into the universality of astrobiology.

This chapter has intended to provide a framework within Darwin's theory of evolution for a preliminary test of the conjecture of the universality of astrobiology. Such a test would be feasible with experiments on the Europan surface, or ocean, involving biomarkers. This aspect of exploration for life in the Solar System should be viewed as a complement to the astronomical approach for the search of evidence of the later stages of the evolutionary pathways towards intelligent behavior (the SETI project).

The set of hypotheses that have been put forward in this chapter are clearly subject to experimental refutation with experiments that are feasible with the current technology that is available to the main space agencies. We have argued that convergence provides a rationale for astrobiology.

10.7 A case for the search for a second Genesis in the Saturn system

Although our understanding of Titan and Enceladus has been greatly enhanced by the data returned by the Cassini-Huygens mission, several aspects related to astrobiology suggest that we should return to the Outer Solar System with more advanced instrumentation. This is a particularly fascinating possibility since the two satellites exhibit atmospheres. Indeed, the scientific community knew the atmosphere of Titan for a long time (cf., Sec. 9.4); but when the Cassini spacecraft approached Enceladus, its magnetometer detected a thin atmosphere. It seems to be the result of venting from ground fractures close to the moon's south pole. The high-resolution images returned by Cassini revealed icy jets coming from the satellite south pole. Life's ingredients, simple organic molecules were detected. The presence of jets implies a warm interior.

The Voyagers showed that Enceladus is only 500 km in diameter and reflects almost 100 percent of the sunlight. Voyager 2 discovered that in spite of the moon's small size, it had a wide range of terrains ranging from old, heavily cratered surfaces to young, tectonically deformed terrain, with some regions with surface ages as young as 100 million years old. Hence Enceladus is the smallest body so far found that seems to have active volcanism. Warm spot in its icy and cracked surface are probably the result of heat from tidal energy like the volcanoes on Jupiter's moon Io, but the exact mechanisms remain a challenge for our future research. Its geologically young surface of water ice, softened by heat from below, resembles areas on Jupiter's moon, Europa.

Other Cassini instrumentation found the atmosphere with about 65 percent water vapor and molecular hydrogen at about 20 percent. The rest is mostly carbon dioxide and some combination of molecular nitrogen and carbon monoxide. The variation of water vapor density with altitude suggests the water vapor may come from a localized source comparable to a geothermal hot spot.

Early in 2008, the Cassini spacecraft performed a flyby through icy water jets that might point to a water ocean inside the little moon. These jets could provide evidence that liquid water is trapped under the icy crust of Enceladus.

As a consequence of these exciting discoveries of Cassini that come more than a decade after the epoch making discoveries of the Jovian moons by the Galileo mission, our general view of the conditions favorable for a second Genesis in our own Solar System, our cosmic backyard, have been greatly enhanced. The conditions for habitability are more widespread than were thought to be possible at the end of the 20th century.

Chapter 11

The Emergence of Intelligence in the Universe

Previous chapters have suggested that a second Genesis could take place in our cosmic neighborhood. But science is in a position to test the emergence of a second Genesis elsewhere in the universe, especially the presence of beings that are able to form a civilization. Indeed, in this book starting from Chapter 2 we have touched upon a possible line of research that is known as bioastronomy. There are evolutionary stages needed for the emergence of intelligence, including the emergence of neurons and brains. These evolutionary events can be documented by exploring the Solar System. In this chapter we shall review some of them.

The reader is advised to refer especially to the following entries in the Glossary: Cenozoic, Hydrothermal vent, Mesozoic, Paleozoic, Prokaryotic, Taxonomy.

11.1 On the inevitability of biological evolution

There is some evidence that once life originates, provided sufficient (geologic) time is available, evolution will provide pressures to living organisms in every conceivable environment. This remark further advocates in favor of the hypothesis that once life appears at a microscopic level in a given planet or satellite, the eventual evolution of intelligent behavior is just a matter of time. Cambrian fauna (existing between 570-510 million years before the present, Myr BP), lamp-shells (inarticulate brachiopods) and primitive mollusks (Monoplacophora), were later maintained during Silurian times (440-408 Myr BP) by microorganisms that lived in hydrothermal vents[1]. Taxonomic analysis of Cenozoic fossils suggests that shelly vent taxa are not ancestors of modern vent mollusks, or brachiopods[2]. We may conclude that modern vent taxa support the hypothesis that the hydrothermal-vent environment is not a "refuge for evolution". In fact, there is evidence that since the Paleozoic (e.g., the Silurian) and through the Mesozoic there has been movement of taxonomic groups in and out of the vent ecosystem through time—no single taxon has been able to escape evolutionary pressures[3]. Hence, these remarks rule out the possibility that even these deep-sea environments are refuges against evolutionary pressures. In other words, the evidence so far does not support the idea that there might be environments on Earth, where ecosystems might escape biological evolution, not even at the very bottom of deep oceans.

These remarks give some support to the hypothesis that any microorganism, in whatever environment on Earth, or elsewhere, would be inexorably subject to evolutionary pressures. As we have shown above, fossils from Silurian hydrothermal-vent fauna demonstrate that there has been extinction of species on these locations, which at first sight seem to be far removed from evolutionary forces. For these reasons we may raise the question whether over geologic time, in the absence of refuges against evolution, the most primitive cellular blueprint (prokaryotic) has bloomed into full eukaryogenesis and beyond, along the evolutionary pathway to organisms displaying intelligent behavior.

11.2 Evolution should be convergent in the cosmos

In order to investigate whether evolution of life is subject to convergence throughout the cosmos, there are at least two options. Firstly, to test directly whether the evolution of intelligent behavior has followed a convergent evolutionary pathway elsewhere in the universe. In principle this is subject to testing by means of the SETI project[4]. Unfortunately, no definite signal that could be interpreted as originating from an advanced civilization has ever been detected. A second alternative, although much more restricted in its scope, is currently in progress: to test for the possible existence of the lowest stages of the evolutionary pathway within the Solar System, namely at the level of microorganisms. One possibility is at present being carried out in the search for life directly on Mars. Subsequently a second possibility within the Solar System was emphasized in the previous chapter, namely the search for life on Europa, the Jovian satellite[5] that will be reconsidered in Sec. 12.4.

Even beyond our own Solar System, scientific research may help us to question whether there are environments that fulfill conditions favorable to life's origin and evolution. This is possible due to the fact that at present we are aware of multiple examples of solar systems. In addition, we suppose that in extra-solar planets steady conditions are preserved. By steady conditions it should be understood that the planet (or satellite) where life may evolve is bound to a star that lasts long enough. In other words, the time available for the origin and evolution of life should be sufficient to allow life itself to evolve before the solar system of the host planet, or satellite, reaches the final stages of stellar evolution, such as the red-giant and supernova phases. It is also assumed that major collisions of large meteorites with the world are infrequent, after the Solar System completed its early period of formation.

Under such steady conditions, the gradual action of natural selection will be expected to be the dominant mechanism in evolution. Fortunately, the existence of steady Earth-like planetary conditions is an empirical question for which we will be able to give partial answers in the foreseeable future. Reliable observational techniques will eventually be provided by progress in space interferometry. In the foreseeable future within the realm of scientific research, we will be able to address the question of the inevitability of the evolution of intelligent behavior in the following sense: evolution of the cosmos, and especially biological evolution right from the earliest stages at the biochemical level, may be "fine-tuned" for the inevitable emergence of intelligent behavior throughout the cosmos, given the assumed universality of biological evolution, and provided steady planetary conditions are preserved in planets or satellites over geologic time.

11.3 Natural selection and convergent evolution

A third factor that militates in favor of the inevitability of the evolution of intelligent behavior in the cosmos is the remark that natural selection seems to be powerful enough to shape terrestrial organisms to similar ends, independent of historical contingency. Likewise, in view of the assumed universality of biology, we would expect similar evolutionary processes to take place elsewhere in the universe. We proceed to discuss some examples that support this view. Before we approach once again the question of convergent evolution, however, we should recall, firstly that the set of factors influencing the relative degree to which the Earth biota has been shaped is still a debatable topic. According to the hypothesis of universal Darwinism, life on Earth, and similarly elsewhere, may have been shaped either by contingency or by the gradual action of natural selection. It may be possible though to document convincingly whether, independent of historical contingency, natural selection is powerful enough for organisms living in similar environments to be shaped to similar ends. For this reason, we highlight the following examples, which suggest that, to a certain extent and in certain conditions, natural selection may be stronger than chance:

• Black European fruit flies (*Drosophila subobscura*) were transported to California over 20 years ago. This event has provided the possibility of testing the role of natural selection in two different continental environments. Pacific coast *D. subobscura* (Santa Barbara to Vancouver) was compared in wing-length with European ones (from Southern Spain to the middle of Denmark). After half a dozen generations were observed in similar conditions, the increase in wing length was almost identical (4 percent). This is a compelling case in favor of the key role played by natural selection[6].

• Anole lizards from some Caribbean islands (*Anolis spp.*) provide another example of evolutionary convergence. The islands are Cuba, Hispaniola (shared by Haiti and the Dominican Republic), Jamaica and Puerto Rico (the so called

Greater Antilles). The observed phenomenon suggests that in similar environments adaptive radiation can overcome historical contingencies in order to produce strikingly similar evolutionary outcomes. We could even say that there has been *replicated adaptive radiation* in the various islands. In fact, what has been shown is that although it had been known that many species thrive on these islands, some groups of lizards from different islands living in similar environments also look similar[7]. Genetic analysis has shown that similar traits have evolved in distantly related species for coping with similar environments (such as tree-tops or ground-dwelling): anoles that live on the ground have long, strong hind legs, while those living at tree tops have large toe-pads and short legs. Repeated evolution of similar groups of species (both morphologically and ecologically) suggests that adaptation is responsible for the predictable evolutionary responses of the anole lizards of the Caribbean. We can speak in this case of evolutionary history repeating itself[8].

11.4 Constraints on convergent evolution

In deciding whether the standard laws of physics and biology imply the evolution of intelligent behavior, it is instructive to appreciate the implications of the existence of several *constraints on chance*. These constraints are relevant to the question of whether life elsewhere might follow pathways analogous to the ones it has followed in the case of terrestrial evolution. Various examples of constraints on chance have been enumerated elsewhere[9-10]:

• Not all genes are equally significant targets for evolution. The genes involved in significant evolutionary steps are few in number; they are the so-called regulatory genes. In these cases mutations may be deleterious and are consequently not fixed.

• Once a given evolutionary change has been retained by natural selection, future changes are severely constrained; for example, once a multicellular body plan has been introduced, future changes are not totally random, as the viability of the organisms narrows down the possibilities. For instance, once the body plan of mammals has been adopted, mutations such as those that are observed in *Drosophila*, which exchange major parts of their body, are excluded. Such fruit-fly mutations are impossible in the more advanced mammalian body plan.

• Not every genetic change retained by natural selection is equally decisive. They may lead more to increasing biodiversity, rather than contributing to a significant change in the course of evolution.

Implicit in Darwin's work we have chance represented by the randomness of mutations in the genetic patrimony, and their necessary filtering by natural selection. Astrobiology forces upon us to accept that randomness is built into the fabric of the living process. Yet, contingency that is represented by the large number of possibilities for evolutionary pathways is limited by a series of constraints. Natural selection seeks solutions for the adaptation of evolving organisms to a relatively limited number of possible environments. From astrochemistry we know that the elements used by the macromolecules of life are ubiquitous in the cosmos.

To sum up, a finite number of environments force upon natural selection a limited number of options for the evolution of organisms. We expect that convergent evolution will occur repeatedly, wherever life arises. It makes sense to search for the analogues of the attributes that we have learnt to recognize on Earth, including the evolution of intelligent behavior. This remark clearly militates in favor of the appropriateness of most searches for a second Genesis, as currently undertaken by the major space agencies. We shall return to this significant remark in the following chapter.

Chapter 12

Intelligibility in the Dialogue Between
Science and Religion

In the present and final two chapters we return to the frontier of astrobiology and the humanities. We attempt to analyze the dialogue between faith and reason that emerges from the possibility of discovering a second Genesis. We attempt to analyze the bases of the dialogue between faith and reason that is between science and religion, in order to demonstrate that some of the more recent aspects of this dialogue have been typical of the continuous evolution of science and monotheistic religions. We will conclude that the current progress of theological and scientific thought, if correctly interpreted and taught in the academic curricula, can only reinforce each other, as both have always striven towards the search for truth[1].

The reader is advised to refer especially to the following entries in the Glossary: Atheism, Deism, Kenosis, Metaphysics, Process philosophy, Process theology, Neodarwinism, Revelation, Viking, Voyagers 1 and 2. He should also read again the entries for Divine action and Positivism.

12.1 The dialogue between science and religion

As pointed out by Barbour both biblical and Greek thought had assumed that nature is intelligible[2]. But a special aspect of intelligibility that the Greeks emphasized from the beginning of philosophy was order. The assumption that order is inherent in nature philosophy was born with the possibility to attempt to rationalize the world that surrounds us. Philosophy provided another significant cornerstone in our search for the intelligibility of nature. If we ask: What is the position of science in contemporary culture? The answer goes from the realm of science to philosophy, where we can search for a rational answer.

Even if a given scientist rejects philosophical thought as being irrelevant to his thinking, he is nevertheless assuming a philosophical doctrine that is subject to discussion. Indeed scientism is the view, well outside the domain of science itself that insists that science can explain all human conditions both mental and physical.

On the other has it has to be underlined and written in bold letters that the great ascendancy and undeniable success of science since the Renaissance with Copernicus, Galileo and Darwin has been due to the fact that the scope of science is both modest and humble. Good science has been built on the shoulders of those giants that succeeded to make accurate, repeatable experiments or observations shared by everyone. Another equally significant cornerstone of scientific success has been to limit speculation only to those theories that were subject to precise formulation, either in terms of mathematical language, or in the case of the evolutionary theories, rationalization of the intelligibility of nature was restricted to exhaustively documented hypothesis of which natural selection stands out as one of the greatest achievements of human reasoning.

In the seminal book *"Out of My Later Years"*[3] Einstein has insisted in some of these fundamental questions that lead naturally to the dialogue between science and religion. Science can only ascertain *what is*, but not *what should be*. Religion on he other hand, deals only with evaluations of human thought and action. Conflicts begin to arise when humanists insist on the absolute truthfulness of all statements recorded in the Holy Books of the Abrahamic religions. At this stage the frontier of science has been crossed. In the same context Einstein makes an interesting statement that explains the relevance of the dialogue between science and religion. He accepts that aspiration to searching towards truth is a prerequisite to create science. This is a feeling that that springs from the sphere of religion. Einstein also raises the central question of the present book, in the sense that faith in the intelligibility of nature is another characteristic of scientists. (We shall return to these inspiring thoughts of Einstein in Sec. 13.4.)

In *"The City of God"*[4] Augustine was influenced by the philosophy of Plato. His writings demonstrate that he was well informed of various branches of knowledge, including mathematics, astronomy and medicine[5]. He depended on his contemporaries that included Pliny, Cicero and the Neo-Platonists. This treatise deals with two cities: the heavenly and terrestrial cities. He interprets the Bible in the context of philosophy. Augustine saw no contradictions in his readings of the Bible (cf., Sec. 3.5).

Other concepts in the humanities are relevant in the question of science and religion: atheism is the disbelief in the existence of God. They may have been led to this view by logical positivism, holding that since assertion about God are incapable of empirical verification they are meaningless. However a different view arose in England soon after the milestone of the theory of evolution was published. Indeed, the Metaphysical Society[6] was an association of that met in London in the decade that followed the publication of Darwin's *"The Origin of Species"*. The Society gathered a very broad range of individuals that included bishops, archbishops, deists, positivists and even atheists. It also included Thomas Henry Huxley, an English biologist, educator and a vigorous supporter of Darwin's evolutionary ideas. His outstanding public statements earned him the nickname "Darwin's bulldog." Rather than accepting to be called an "atheist", he preferred to be called an "agnostic", who did not deny or affirm God's existence. He mainly advocated a positivist position that is shared by contemporary science, in which biology and other sciences dealt only with the knowable world (cf., Chapter 2, Sec. 2.5). In fact, today we understand by agnosticism the doctrine that humans cannot know of the existence of anything beyond the phenomena of their experience. The term is sometimes used loosely implying skepticism about religious matters. The term has come to be equated skepticism about religious questions in general and in particular with the rejection of traditional Christian beliefs. This doctrine has been widespread in Western civilization. There are some related and significant philosophical and theological issues to discuss arising from the insights into our origins that have been raised in this book.

The question *Is truth relative?* leads to yet another term form the humanities that is needed to understand the full implications of the dialogue of science and religion that includes the general philosophical topic of relativism[7,8]. According to this doctrine we no longer seem to have any absolute values. It would even be argued that modern thinking should be adjusted to incorporate relativism.

Against the background of the various views of humanists mentioned above, astrobiology aims at the larger questions of modern science, while being squarely set on scientific and technological tools. Science is searching a second Genesis. As we asked at the beginning of this section for science in general: What is the position of astrobiology in contemporary culture? The answer should be searched in the realm of philosophy can be appealed to for a rational answer. But in this book we focus on the scientific question of whether there is a second Genesis in our Solar System.

12.2 Is there a second Genesis in our Solar System?

The question of whether life has emerged in the Solar System was first addressed successfully on Mars with two identical spacecrafts, each consisting of a lander and an orbiter. The Vikings 1 landers touched down on different Martian hemispheres[9]: Viking 1 landed on Chryse Planitia, while Viking 2 landed at Utopia Planitia. The two landers conducted three biology experiments designed to look for possible biomarkers. These experiments provided no clear evidence for the presence of living microorganisms in soil near the landing sites, although the interpretation of the original experiments has continued over the years. The presence of solar radiation in the ultraviolet, the extreme dryness of the soil and the soil chemistry are factors that prevent the formation of living organisms in the Martian soil. This does not preclude that in the remote past conditions may not have been more favorable to the origin of life. In fact, most of the current fleet of space missions with Mars as their main objective includes measurements of a geologic nature that could provide some hints of he possible habitability of Mars 3 to 4 billion years ago.

The Viking Orbiter 1 continued for four years, concluding its mission August 7, 1980, while Viking Orbiter 2 functioned until July 25, 1978. Viking Lander 1 made its final transmission to Earth in 1982. Data from Viking Lander 2 arrived at Earth two years earlier. The Voyagers are two US spacecrafts that explored the outer Solar System after the Viking probes during the period 1977-1989. Data and photographs transmitted by the Voyagers revealed previously unknown details about each of the giant planets and their moons. Close-up images from the spacecraft uncovered a variety of phenomena in the Jupiter system Jupiter and volcanic activity on Io, one of its so-called Galilean satellites. Amongst the Voyager-discoveries one of the most significant was related to the second of the satellites that was discovered by Galileo in 1610: the satellite Europa is a possible body where a second Genesis may have taken place in the Solar System. This follows from the indication that a large proportion of the detectable material on its surface is water[10]. The two Voyagers were launched the same year, but their objectives were different. Voyager 2 completed a set of images of every planet with the exception of Pluto, a project that will be completed in 2015 by the NASA New Horizons Mission. According to the results obtained on the Jupiter system itself by Voyager 2, Europa is covered by a layer of ice under which there may be an ocean of water, at temperature of 4°C. The excitement surrounding Europa is due to the Galileo mission (1995-2003). This mission has changed the way we look at the Solar System and especially Europa. This mission was the first to conduct long-term observations of the Jovian system from orbit. It found evidence of subsurface saltwater on Europa, Ganymede and Callisto and revealed the intensity of volcanic activity on Io. From the similarity of the processes that gave rise to planets and satellites, we may expect that hot springs may lie at the bottom of the ocean. It has been assumed in the past that Jupiter's proto-nebula must have contained many organic compounds. Organisms similar to heat-loving microbes could possibly exist at the bottom of Europa's ocean. However, given the incomplete understanding of the evolution of early life on Earth, at present we should allow microorganisms as a possible Europan biota.

We may add that up to the present time we do not fully understand the evolution of the earliest ancestor of all life on Earth. Indeed, plate tectonics has obliterated fossils of early organisms from the crust of the Earth, which is the only record available to us the evolution of early life.

In passing, it is interesting to remark that Voyager 2 continues to contribute to our understanding of the Solar System: the beginning of the transition zone between the heliosphere (the solar wind bubble, cf., Sec. 7.5) and the rest of interstellar space is known as the "termination shock". In the year 2008 Voyager 2 has crossed this boundary closer to the Sun than expected. This suggests that the heliosphere in this region is pushed inward, closer to the Sun, by an interstellar magnetic field[11].

12.3 The possibility of the emergence of life on Europa

Of the above-mentioned two forms of heating, on Earth we are familiar with radiogenic heating, which is a consequence of the heat produced by the accumulation within the Earth crust of radioactive compounds. The new factor in the Jupiter system is that unlike the Earth, Europa is influenced by its two neighboring satellites: the very volcanic satellite Io, which is closer to Jupiter and the giant satellite Ganymede, which is even larger than planet Mercury. The dynamics of this three-body system keeps the satellites from perfect circular orbits. The consequence is that the eccentricity of the Europan orbit varies significantly in the points of closest and farthest approach to the giant planet, thereby creating thermal gradients that we have called earlier "tidal heating".

Fortunately, we have some clear observational evidence of this form of heating, since Io, its nearest Moon-sized neighbor, is not covered with ice like Europa itself, neither has it a thick atmosphere such as that of Titan, the satellite of Saturn. It is possible then that the Europan internal ocean has consequently been formed (underneath a relatively thin ice cover) through dehydration of silicates, the heating source being due to tidal heating with an addition due to radiogenic heating. For the above reasons it was estimated that the temperature underneath the icy crust could be 4°C.

From the similarity of the processes that gave rise to the solid bodies of the Solar System, we may expect that hot springs may lie at the bottom of the ocean. The main thesis of the proponents of the existence of an Europan biota is that, as Jupiter's primordial nebula must have contained many organic compounds, then possibly, organisms similar to heat-loving microbes can evolve at the bottom of Europa's ocean. The previous argument correctly pointed out that the most important requirements for the maintenance of life in Europa are the above-mentioned conditions (liquid water, an energy source and organic compounds).

12.4 Future missions to Europa

The exploration of the Jovian System and its fascinating satellite Europa is one of the priorities considered by the main space agencies. The Jovian System indeed displays many facets. It is a small planetary system in its own right, built-up out of the mixture of gas and icy material that was present in the external region of the solar nebula. Unique among Jupiter's satellites, Europa is believed to shelter an ocean between its active icy crust and its silicate mantle, one where the main conditions for habitability may be fulfilled. In order to decide whether Europa is really habitable, we need a dedicated mission to Europa. There are missions that are being considered by the main space agencies. In one of them system a triad of orbiting platforms will be deployed on the icy surface, in order to perform coordinated observations of its chemical and physical properties. One of the highest priorities of all future missions will be the potential habitability of Europa. This property rests on the fulfillment of four conditions: the presence of liquid water, an adequate energy source to sustain the necessary metabolic reactions, a source of the biogenic elements (C, N, H, O, P, S), which can be used as nutrients for the synthesis of biomolecules, and relevant pressure and temperature conditions[12].

Europa is not unique among the four Galilean satellites, as subsurface oceans may also exist at Ganymede and Callisto. But according to current models, it represents the only case in which liquid water is in contact with a silicate core. Such conditions are favorable for interactions between the ocean and silicates, particularly if a volcanic, and consequently hydrothermal activity exists. This would provide a variety of chemicals that could play a role in sustaining putative life forms at the ocean floor.

12.5 On the implications of Darwinism

The search for life on Europa represents one of the major efforts that will have to be faced in the foreseeable future to give us insights into life's origin that would go beyond what we have been able to learn from research in organic chemistry. The general outline of life's emergence on Earth is nevertheless understood, and its implications for the humanities have been discussed elsewhere[13]. For example, life is known to have emerged early in the history of the Earth, some 3.5 billion years before the present. The evidence comes form micropaleontology[14]. The microorganisms involved are fossils of stromatolites. These are geological features consisting of a stratified rock formation, which are essentially the fossil remains of bacterial mats. The bacteria that gave rise to these formations were mainly cyanobacteria. Similar mat-building communities can develop analogous structures of various shapes and sizes in the world today.

Besides, it is safe to assume that over 3 billion years ago there was a flora of cyanobacteria, although the exact date for the earliest ancestor is a hotly debated issue[15]. The early forms of life are also known to have been a major factor in the evolution of the hydrosphere and the atmosphere. The time sequence of these events has been inferred following standard procedures and hypothesis. Like other branches of science, these dates are subject to improvements by new experimental techniques and observation procedures. This state of affairs is in sharp contrast with the questions of faith that are based on tradition and revelation. The statements of science are closer philosophy in the sense that both systems attempt to base their statements on rational basis. As we have seen in previous chapters, Darwin's work led to the definitive theory of evolution that had been anticipated in earlier incomplete forms by Charles' grandfather Erasmus Darwin and independently by Jean-Baptiste Lamarck.

The work of Lamark that is especially relevant in the present context. He published in 1778 a successful book on French plants to great acclaim. This academic success led to his appointment as an assistant botanist at the royal botanical garden that was not only a botanical garden but also a center for biological research. At the beginning of the 19th century Lamarck in his late 50s began the revolutionary steps that led him to develop an evolutionary theory, rather than accepting that the living world was fixed and harmoniously organized. Lamarck took on the challenge of a new field of biology. The diversity of invertebrates proved to be a rich source of knowledge. Lamarck believed that a change in the environment causes changes in the needs of organisms living in that environment, which in turn causes changes in their behavior, and once again in turn this leads to differential use of internal organs. (Eventually organs either continuously improve or are lead to their gradual disappearance.) The mechanism for the evolution of life on Earth (Lamarckian evolution) is different from that proposed by Charles Darwin, however, the predicted result is the same, namely adaptive change in lineages, ultimately driven by environmental change, over long periods of time. In spite of its limitations, the originality of evolution of life on Earth being driven by the environmental changes introduced a new point of view in the life sciences. According to Lamarck living species are inter-related through reproduction, slowly evolving through the course of generations.

Throughout his life Darwin avoided the problem of the origin of life, except from making a few speculative remarks about the possible environments where life could have originated, the so-often-quoted "warm little pond". This was a very reasonable attitude for the late 19th century, before experimental science began to address the question of the origin of life.

12.6 Darwinism, philosophy and theology

The rationalization of a natural history in terms of Darwinism has been argued to render the intelligibility of nature more plausible in terms of theological arguments. Darwin focused on the origin of the species by introducing the term "natural selection" for reproductive success, allowing adaptation to changing environmental conditions. In other words, natural selection is the non-random element in evolution that gives evolution its direction. The magnitude of this revolutionary contribution to science is evidenced by the perennial difficulty to insert this aspect of science into the mainstream of cultural knowledge. This may be illustrated with the dialogue between his Eminence Cardinal Christoph Schönborn and George Coyne SJ[16,17].

The main point made by Schönborn is that evolution in the sense of common ancestry might be true, but evolution in the neo-Darwinian sense—an unguided, unplanned process of random variation and natural selection—is not. Any system of thought that denies or seeks to explain away the overwhelming evidence for design in biology is ideology, not science.

However, the shared by most scientists[18] is that science is completely neutral with respect to philosophical or theological implications that may be drawn from its conclusions. Those conclusions are always subject to improvement. As we have emphasized above this is the nature of science. But to deny today's science on religious grounds is to go beyond the natural boundaries of theological thoughts. Likewise, to attempt to make changes in theological thought on the strength of science is to go beyond the natural boundaries of science. Western civilization has faced this dichotomy before during the Enlightenment. (cf., Chapter 2, Sec. 2.5).

We have seen that this group of philosophers maintained that scientific knowledge is the only kind of factual knowledge. Although some scientists have adopted this philosophy, either consciously or unconsciously, the fact remains that modern science begins with Galileo, who initiated the tradition of formulating theories based on observation and experiments. No underlying philosophy was adopted then, or need to be adopted now, beyond the dialogue between theory and experiment.

Positivism avoided all considerations of ultimate issues, including those of metaphysics and religion. However, the reduction of all knowledge to science is a matter that debate has not yet settled. Natural selection was expanded by the gradual growth of the science of genetics, especially molecular genetics. The origin of life on Earth is on similar grounds; the phenomenon is understood in its broad outline.

The details of the specific organic pathway from biochemistry to a microorganism are still in being searched. But our theories imply that once the biomolecules of life were self-assembled into a functional living cell, natural selection led to evolutionary pathways that left clear records in the fossils.

12.7 On the progress of the different areas of culture

So far the above discussion of the emergence of life on Earth is subject to experimental refutation. The example mentioned earlier in relation to dating of the oldest microfossil of 3.5 billion years before the present is a clear example of the way science progresses. Refutation of arguments previously assumed to be on secure basis constitutes a normal procedure in science. The philosophical discourse, on the other hand, is in a "no-man's land" between science and religion. One clear illustration is provided by the question of relativism. One universal truth in a theological sense is the omnipresence of Divine action, which is an act of faith that requires no scientific support, neither is it subject of the relativist constraints. As a question in the philosophy of religion, relativism may be marginally a more interesting topic. But relativism is irrelevant to faith that is based on tradition and revelation. These premises of faith are common to the whole Abrahamic tradition that is professed by Jews, Christians and Muslims alike.

We may conclude by saying that the distinct pathways that science and faith follow are quite independent. With Augustine we believe that it is erroneous to attempt to confront the scientific implications of natural selection with religious statements from the Holy Books (cf., Sec. 9.5). This would take us beyond the natural frontiers of science. It would be equally erroneous to confront the theological statements related to Divine action with scientific concepts for this would break the boundary of theological thinking that should only concern revealed truth and related rational thinking.

No controversy should arise in relation to teaching science and religion. If natural selection were eventually faced with scientific facts and observations that suggest it not to be the best hypothesis for the evolution of life on Earth, then a better mechanism would be suggested, due to newer facts that would not allow a traditional Darwinian interpretation.

Likewise if new prophecies were to be lead to new revealed truth in the realm of theological thought, then this would lead to a new theologies, not to new scientific approaches. This has occurred repeatedly in Western civilization when, for instance, the revealed truth of Judaism was supplemented with the revealed truth of Christianity, and later by that of Islam. Nevertheless, these parallel avenues of thought often address the same questions, such as what is the relation between humans and the universe.

However, before abandoning this topic, it is important to underline that Darwinism and theological thought have been shown to be compatible within the framework of kenotic process theology. Indeed, kenosis is concept is taken to mean self-emptying and voluntary sacrifice on behalf of others, based on genuine and freely given love for others, and resulting in generosity and respect that flow from it. Haught has emphasized God as the sole ground for the world's being[19, 20]. This approach to natural theology leads him to explain the world in terms of evolution, as understood within the Darwinian tradition[21]. He focuses on features of process thought.

This philosophical system is considered to be particularly helpful in the task of constructing an evolutionary theology that may throw some further insights on Darwinism. To sum up, we have attempted to analyze the bases of the dialogue between the new science of astrobiology and religion, in order to demonstrate that some of its more recent aspects of this dialogue have repeatedly accompanied the progress of science and the progress of monotheistic religious thought. We conclude that the current considerable progress of theological and scientific thought, if correctly interpreted and taught in the academic sector, can only reinforce each other, as both have always strived towards the search for truth.

Chapter 13

The Ultimate Frontier of Science and the Humanities

We consider questions of faith and reason that arise from the origin and evolution of life on Earth. This common objective—to reflect on the emergence of life on Earth from two different cultural points of view—has led in the past to a fecund dialogue between faith and reason.

The reader is advised to refer especially to the following entries in the Glossary: Einstein, Ethics, Leibniz, Neoplatonism, Pantheism, Plato and Theism.

13.1 A fruitful dialogue between science and religion

We face the dialogue between science and religion with the conviction that it will be fruitful, because the search for the truth is a common objective of both of these aspects of culture[1].

We are also convinced of the unity of knowledge. A fractured unity is due to the current excessive specialization that puts a barrier to the eventual integration of science and religion in our society. Narrow expertise may distance culture from the bigger picture that is relevant for unified knowledge of vital importance for discussing the question of faith and reason, especially regarding Divine action and evolution by natural selection.

Profound changes have taken place in the understanding the constraints imposed by philosophy and natural theology on our view of life. They mostly favor a special place for man in the universe. Neo-Platonism is a philosophical doctrine. Plotinus, who died in 270 AD, is often considered to be its founder. He was an expositor of the doctrines of Plato. This philosophic school drew not only upon the Platonic canon, but also upon Pre-Socratic literature. We recall that Platonism is the doctrine that asserts the existence of such things as abstract objects and, hence, they are entirely non-physical and non-mental. With Marsilio Ficino (1433-1499) we may say that Platonism achieved a brief moment of ancient renewal, when this philosopher advocated a special place of man in the universe. On the other hand, within the boundaries of science with its characteristic experimental and observational methods, a different outlook has been inferred. The general laws of science do not assign humans any special position in our galaxy. Science since Copernicus gives our nearest star, the Sun, an insignificant position in the cosmos.

13.2 Frontiers of the humanities

Science has made enormous progress since Galileo's epoch-making strict adoption of the experimental approach for understanding nature. Modern science since Galileo recognizes that scientific ideas must be provisional and capable of being overturned by evidence from experimentation and by observation, which is its most robust and strong feature. There appears to exist a misunderstanding in relation with a universe evolving for 13.7 billion years since the Big Bang. In the cosmos, beginning at about 3 to 4 billion years before the present life evolved, in its most primitive forms, through a process of random genetic mutations and natural selection. Science is neutral with respect to philosophical or theological implications that may be drawn from its conclusions. Those conclusions are always subject to improvement. Indeed, the National Academy of Sciences has expressed this key issue that is significant for scientists that are also believers, in very clear terms[2]:

Many religious persons, including many scientists, hold that God created the universe and the various processes driving physical and biological evolution and that these processes then resulted in the creation of galaxies, our solar system, and life on Earth. This belief, which sometimes is termed "theistic evolution", is not in disagreement with scientific explanations of evolution. Indeed, it reflects the remarkable and inspiring character of the physical universe revealed by cosmology, paleontology, molecular biology, and many other scientific disciplines.

Religion has come a long way in its natural-theological approach to deeper philosophical questions, always basing its considerations on revelation and tradition. Both scientists and humanists are well represented in this book with their worthy reflections.

It is understandable that in addressing the phenomenon of the origin and evolution of life in the universe, the ensuing dialogue has not always excluded mutual contradictions. These discussions are necessary to delineate clearly the frontiers of each contribution. In this respect we should not forget Galileo's statement that the Book of Scripture and the Book of Nature speak of the same God. In a modest effort towards aiming at the unity of knowledge we have included many contributions in these pages. We have brought together multiple points of view. Heterogeneous in their approaches from either humanities or science, the chapters have been brought together, always keeping in mind the strict frontiers of either theology, or science.

In the present context, one of the most distinguished philosophers of modern times, Bertrand Russell, explained this point particularly well. Lord Russell (1872-1970) was an English logician and philosopher, whose seminal work in mathematical logic was published in the early 20th century. Russell collaborated with Alfred North Whitehead on *"Principia Mathematica"* (1910-1913). Russell's interests ranged over a wide spectrum including philosophy, mathematics, science, ethics, sociology, education, and history. And most significantly from the point of view of this book his interests also extended to the area of religion. He had a personal gift in explaining philosophical and scientific arguments, an unusual ability that led to his 1950 Nobel Prize in literature[3]. His main point (quoted in full at the beginning of Chapter 1) is that philosophy is something intermediate between theology and science. Like theology it consists of speculations on matters as to which definite knowledge has, so far, been unascertainable. But Russell clarifies that like science, philosophy appeals to reason, rather than tradition, or revelation. He underlines that philosophy is like a "no-man's land" that is approachable from science, and also from theology. A common denominator amongst the disagreements presented in this book is still missing a clear demarcation of the frontiers of science. In other words, the three areas demarcated by Russell are not always maintained. To a large extent these disagreements reflect present-day uncertainties and a failure to popularize science adequately.

13.3 Questions from the philosophical area of ethics

Another important topic of faith and reason that is pertinent to our arguments concerns a certain caution of theologians with respect to the questions on which we have dwelt at some length. As we have mentioned earlier in the book the Bible is not a single book, but rather a library with texts of different nature, but uniformly displaying a deep theological content revealing some aspects of God and his Divine action. Today, bearing in mind the significant progress in the space sciences, especially in astronomy, a quotation from the Book of Daniel is particularly relevant. The Book of Daniel is a book both in the Hebrew Bible and the Christian Old Testament. The book revolves around the figure of Daniel, an Israelite who becomes an adviser to Nebuchadnezzar, the ruler of Babylon from 605-562 BC. The book may have been written during Daniel's lifetime, or it may have come to us from a version written later containing an inspiring poetical analogy with the stars in the firmament comes from Daniel 12: 1-3:

> *And those who are wise shall shine like the brightness of the firmament; and those who turn many to righteousness, like the stars forever and ever.*

This quotation should be taken strictly as an analogy, and not as scientific information on the longevity of stars, since we know today from a considerable amount of data on supernovae that stars are not eternal, but once their nuclear fuel runs out several scenarios are possible, including the supernova explosions. But the message in the Book of Daniel is valid today. It still retains its usefulness after well over two millennia since the prophet's words were written. Daniel's main point is not one of accuracy in astronomy, but rather it is a point that emphasizes the importance of good living and good behavior.

To such questions pertaining to the philosophical area of ethics, no answer can be found in the laboratory, and the prophet's comments are relevant. Additional discussions on the question of faith and reason are available[4-12]. Besides, books have contributed to these discussions[13-15]. Today the subject of faith and reason has captured the attention of the general public.

13.4 Views on science and religion

Albert Einstein, Charles Darwin, Isaac Newton and Nicholas Copernicus are amongst a handful of scientists who have radically transformed our view of the world. In particular, Einstein considered that a conflict arises when a religious community insists on the absolute truthfulness of all statements recorded in the Bible[16]. He had an impersonal view of God. He believed, in agreement with Spinoza, that everything that exists is God (a doctrine known as pantheism). But he opposed the view that God is no more than the sum of what exists, since He had infinite qualities. Only two qualities, thought and extension can be perceived by human intelligence. Hence, God must also exist in dimensions far beyond those of the visible world[17].

Indeed, the literal interpretation of the Bible means an intervention on the part of religion into the sphere of science. What is most significant over half a century after Einstein wrote these words are Einstein's comments on the relation between faith and reason. Scientists have an analogous faith that the world is intelligible to reason in terms of a mathematical description of the physical sciences. Einstein concludes this part of his reflections in "*Out of My Later Years*" with the statement that he cannot conceive of a genuine scientist without that profound faith. It is at this stage that Einstein inserted his often quoted opinion:

Science without religion is lame, religion without science is blind.

13.5 Further insights into our origins

As we have learnt throughout this book, the new science of astrobiology covers research in the field of the origin, evolution, distribution and destiny of life in the universe. Astrobiology is currently in a period of fast development due to the many space missions that are in their planning stages, or indeed already in operation[18].

However, even though no single living cell has been formed as yet in the test tube, chemical evolution, as we have attempted to demonstrate in previous chapters, continues to be a solid scientific pursuit that has been reviewed extensively over the last decade[15,19-22].

Going beyond Darwinism strictly within the life sciences, including astrobiology is not new. One illustration is provided by adaptive radiation. This is a process in which one species gives rise to multiple species that exploit different habitats that they may occupy. This is an evolutionary process driven by mutation and natural selection. A well-known example of adaptive radiation as the result of an environmental change is the relatively rapid spread and development of mammalian species after the extinction of the dinosaurs. The speed of the adaptation is measured in time scales that are familiar to geologists. For example, the periods of time that are relevant for adaptive radiation are measured in millions of years. Darwin understood such a phenomenon. By studying the work of his contemporaries he brought this phenomenon to the attention of science in his seminal work "*The Origin of Species*"[23].

What is new to us, at the present time is a phenomenon observed in, for example, fish. A small population of Trinidadian guppies was scooped from a waterfall pool, where predators were abundant. Later they were released upstream in a pool in the presence of only one enemy species. The guppies adjusted to the new environment with few predators by growing bigger, living longer, and having fewer and bigger offspring. Although natural selection was assumed by Darwin to be a slow process, it was found that natural selection is able to act speedily, not in the span of millions of years as in the case of the mammalian radiation that took place about 65 million years ago[24]. On the contrary, the guppies adapted to their new environment in a mere four years, which is a rate of change some 10,000 to 10 million times faster than the average rates determined from the fossil record [25].

If Darwinian theory is thought of as being incomplete and we do keep within the boundaries of science, as pointed out above by Einstein, then the evident avenue to follow is to replace the theory by another one that may explain some of the data better. One significant example is already provided by the well-known theory of evolution by "punctuated equilibria". This theory suggests that species are fixed most of the time, and only change when new opportunities open up, usually following the extinction of other species. Darwin on the other hand, appreciated the evidence suggesting that species are fairly stable entities with distinct beginnings and ends. His discovery of natural selection led him to believe that evolution must be slow and gradual[26].

To sum up, if neither Darwinism, nor punctuated equilibrium eventually were shown to satisfy in the future all the data that will be available in the life sciences, then science will show its traditional strength by challenging biologists to produce a more refined theory for the evolution of life on Earth.

13.6 Darwinism within natural theology

Religion has had great difficulty in assimilating the real significance of Darwinism. From the point of view of theology the difficulty focuses on how to reconcile evolution with the idea of Divine action. It is possible to look at the natural world for explanations with scientific ideas that by the very definition of science must be provisional, namely capable of being discarded by evidence from experiments and observation. This approach to science is in sharp contrast intelligent design that will be considered in the next section.

Some scientists say simply that science and religion are two separate realms, "non-overlapping magisteria"[27]. In this view, science is relevant in the realm of "what the universe is made of (fact) and why does it work this way (theory)", while religion concerns itself with questions of ultimate meaning and moral value. As the careful reader will notice, this position is compatible with the above-mentioned Russellian statement on the distinction of the different components of culture.

Besides, in Gould's book, there are various other efforts along the lines of trying to understand the significance of Darwinism in the context of natural theology. In Sec. 12.7 we have brought to the attention of the reader examples of such theologies. For this purpose the concept of kenosis was added to the natural theology to explain the world in terms of Darwinism. It focused on features of process thought.

As we have seen, this philosophical system is considered to be particularly helpful in the task of constructing an evolutionary theology that may throw some further insights on Darwinism[8]. Another approach along these lines points out that theologians already have the concept of God's continuous creation with which to explore the implications of modern science for religious belief[28]. In this view God is working with the universe. The universe has a certain vitality of its own like a child does. It has the ability to respond to words of endearment and encouragement. You discipline a child but you try to preserve and enrich the individual character of the child and its own passion for life. A parent must allow the child to grow into adulthood, to come to make its own choices, to go on its own way in life.

Words that give life are richer than mere commands or information. In such wise ways we might imagine that God deals with the universe. Along the above lines it is possible that evolution may also provide a way in which the tradition of natural theology may undergo a renewal. Instead of focusing design without a designer, which can be accounted for scientifically in terms of Darwinism[29], a revived natural theology may take place if we interpret correctly the origin and evolution of life on Earth.

13.7 Beyond the natural boundaries of science

In spite of the significant opinions expressed by the theologian Saint Augustine, the philosopher Lord Russell and the scientist Albert Einstein that we have mentioned in this chapter, the literal biblical interpretation of the origin of life on Earth by a Divine Designer continues to be discussed.

We have maintained throughout this book that the dialogue between science and theology is fruitful: the search for the truth is a common objective of all aspects of our culture. The unity of knowledge should be preserved. The present fractured unity may be due to the current specialization of professional scientists, philosophers and theologians. Narrow specialization hides the insights that are relevant for a unified knowledge. Avoiding this current state of our culture is of vital importance for discussing the question of science and religion, especially regarding Divine action and Darwinism.

This approach beyond the natural boundaries for science to the emergence of natural events has been called intelligent design (ID). The advocates of ID, outside the natural boundaries of the science of biology, dispute the idea that natural selection fully explains the complexity of life. This represents an unprecedented metaphysical basis for denying the validity of natural selection without providing new experimental, or repeatable observational data.

Besides, ID proponents say that life is so intricate that only a powerful guiding force, or intelligent designer, could have created it. (In fact, science is neutral with respect to this theological statement.) ID does not identify the designer, but ID is one interpretation for God and its divine action in the universe that lies well beyond the realm of science and, most significantly, ID also lies well beyond natural theology, as natural theology is, strictly speaking, the body of knowledge about religion that can be obtained by human *reason* alone, *without* appealing to revelation. It is especially relevant to highlight the fact that ID is in no way whatsoever the only possible approach to gaining insights into divine action in a theological context.

The ongoing debate between Darwinism and ID has been taken back to the school curricula, where it has caused once again an embroiled hostility analogous to the 1925 Scopes Evolution Trial. It was also called the "monkey trial" as a reference to the second aspect of the theory of evolution, besides natural selection, namely the descent from a common ancestor, which at the time of the publication of "*The Origin of Species*", was emphasizing the evolution of primates (and the popular press singled out the word "monkeys"). The general dialogue at the time did not refer to a last common ancestor that astrobiology has traced back to a microbial organism (by careful arguments from molecular biology).

We believe that there should not be a real conflict between the holy books and science. Disagreements can only be apparent through lack of appreciation of the real limits of science and the true nature of natural theology. Judaism and other monotheistic religions may accept natural selection.

13.8 Some questions may escape the scope of science

Stellar evolution predicts that the lifetime of the Sun will continue for a few billion years, while the science of anthropology suggests that humans are a fairly recent addition to an Earth biota, whose origins must be traced back to microorganisms that emerged on the early Earth from 3 to 4 billion years before the present. We expect that theology will continue its independent search for a deeper understanding of God and the meaning of his Divine action. Science will be confronted with ever increasing mysteries, as our scientific instruments become better and more accurate to allow confrontation of theory with experiment. The questions of how the universe and life in it were formed will reveal new insights. Some questions escape the scope of science. We expect that future developments in philosophy and theology will gradually give us better insights into the question that was raised by the philosopher Gottfried Wilhelm Leibniz (1646-1716)[30]:

> *"Why is there something rather than nothing?".... Further, assuming that things must exist, it must be possible to give a reason why they should exist as they do and not otherwise* (Leibniz, 1940).

We are convinced that in the future the question of the proper place of Darwinism in science, and its teaching in colleges, will be settled to the satisfaction of society in general. This will take place without excluding the questions of faith in God that many scientists accept.

It is of paramount importance for non-scientists to understand that we do not consider science as the ultimate truth, but instead, since the time of Galileo Galilei the scientific approach to the intelligibility of nature is one of the great achievements of our culture. Science is a robust search for the truth through accurate measurements and observations that will later receive a theoretical presentation in terms of hypothesis. One excellent example is natural selection— the backbone of the science of biology—that is thoroughly supported by confrontation with natural history observations and data from molecular genetics. Alternatively, accurate measurements and observations lead in sciences such as physics and the space sciences to theoretical descriptions in the language of nature-mathematics. But it is built into the scientific method the need to question current theories with ever increasing accuracy to replace them with more general descriptions that will take into account the new experimental data or accurate observations. The generalization of Newton's theory of gravitation by Einstein's theory of gravitation (General Relativity) is a classical example.

To sum up, we have aimed at stressing that science is an activity whose main strength is that it invites arguments to prove that current theories may be improved; but such proofs are not implemented by acts of faith, or by authority. Progress is encouraged by science being continually corrected with new, relevant data, and by observation with instruments that the vertiginous progress of modern technology has put in our hands.

Chapter 14

Can Nature be Intelligible?

Throughout this book we have provided insights into the intelligibility of nature. We have followed some stepping-stones to show that an interdisciplinary exchange is profitable for the whole of human culture. In this chapter, we summarize aspects of the intelligibility of nature mentioned earlier, namely the origin and evolution of the universe, and the subsequent origin of life. Intelligibility, as we have formulated it in this book, does not end with a preliminary understanding of the emergence of life, but is linked with the evolution of life. The main thread that has run throughout the book has been the possibility of identifying a second Genesis inside the Solar System. Such an achievement will provide significant additions to our insights on the intelligibility of nature.

The reader is advised to refer especially to the following entries in the Glossary: Biogeochemistry, Camus, De Duve, Dirac, Existentialism, Heidegger, Rationalism, Empiricism, Idealism, Reductionism, Salam, Sartre, Schrödinger and Weinberg. He is also recommended to read again the entry for "Logical positivism".

14.1 Intelligibility of nature: a cultural problem

In previous chapters we have developed the following idea in detail[1]: amongst the priorities of the new science of astrobiology, we should consider the search for other lines of biological evolution elsewhere in the universe, a search that we have referred to as a second Genesis[2]. The implications of a second Genesis are likely to be relevant, not only to astrobiology, but also ethics and philosophy. Intelligible means the capability of being understood, or comprehended. Alternatively, intelligible can signify to be apprehensible by the intellect alone. A third aspect of intelligible, closer to the significance of the term in the context of the present work, is related to something that is beyond perception. An "intelligible universe" can be the starting point of a prolonged and systematic discussion amongst astrobiologists, as well as philosophers and theologians. The Belgian Nobel Laureate Christian de Duve, at the end of a review on the origin and evolution of life, asks himself the question: *"What does it all mean?"* Not only in science[3], but also in philosophy[4] and in theology [5] the intelligibility of the universe raises questions that lie on the frontier between science and the humanities. Often considerations of intelligibility return to the often quoted, but less frequently debated statement of Steven Weinberg, the American Physics Nobel Laureate[6]:

The more the universe seems comprehensible, the more it also seems pointless.

We should clarify whether the quoted statement reflects a specific philosophical trend, or attitude, characteristic of the first half of last century. Hence, if that were the case, the statement should not discourage the general dialogue at the frontier of science and religion. The common approach from both ends of the academic discourse to discovering a second Genesis anywhere else in the universe should contribute, as we shall see below, to the progress of the philosophy of religion. Reciprocally, the questions raised in any field of the humanities arising from the discovery of a second Genesis could, in turn, enrich the search for the place of humans in the universe.

The above quotation can be best understood in the context of a philosophical trend called Existentialism. Earlier doctrines (Rationalism, Empiricism and Idealism) arose from the increased scientific knowledge of the 17th and 18th centuries. They had maintained that the cosmos is a well determined ordered system, and hence intelligible to all observers. In that framework there was no motivation for viewing the origin and evolution of the universe and life as being absurd, or pointless, a trend that was to arise some 200 years later. On the other hand, the existentialists went beyond Rationalism that was represented, amongst others, by Benedict De Spinoza the Dutch-Jewish philosopher and foremost exponent of this 17th-century doctrine[7]. The existentialists also went beyond Empiricism. Another influential philosophical view stresses the central role of the ideal, or the spiritual, in the interpretation of experience. Known as Idealism, this philosophic doctrine, unlike Rationalism and Empiricism, maintains that the world, or reality, exists essentially as spirit or consciousness, that abstractions and laws are more fundamental in reality than sensory things. Rationalists, empiricists and idealists laid down solid bases on which to discuss questions related to the eventual destiny of life and of the universe. Rationalists and empiricists had argued that we could discover all natural universal laws by reason and experimentation, largely in agreement with the emergence of modern science with Copernicus, Galileo, Bruno, Digges, Newton and others.

A systematic idealist, Georg Hegel attempted to elaborate a comprehensive and systematic ontology from a logical starting point[8]. In other words, although differing somewhat from other idealists Hegel attempted to defend faith as being logical. This movement was influenced by the growth of science since the Enlightenment. The need was felt for a harmonious development of human culture.

The bases of the philosophy of religion had to be extended to accommodate so much new revolutionary scientific knowledge. Faith in a Creator had to be seen in a new light. The Danish philosopher and religious writer Soren Kierkegaard opposed Hegel's views, particularly the concept that religious faith was logical; he explores the notion of the absurd: Abraham gets a reprieve from having to sacrifice Isaac, by virtue of the absurd[9]. He further anticipates the still-to-come philosophical trend (Existentialism), insisting in a thesis opposite to Hegel's: humans suffer a deep anxiety (and hence need religion) because one has no certainties[10]. In more modern terms we can paraphrase Kierkegaard and other pioneers of Existentialism by saying that life in the universe is pointless and absurd. To put it simply, according to Kierkegaard, life in the universe is not intelligible. Consequently, the questions of our destinies could not be incorporated as a frontier between astrobiology and the humanities that would encourage a fruitful dialogue. A third relevant aspect of Existentialism came from the implications of the philosophy of Martin Heidegger[11], as interpreted by his close follower Jean Paul Sartre[12]. Heidegger formalized Existentialism on the basis of the work of Edmund Husserl, a German philosopher who is well known as the founder of Phenomenology[13]. This earlier method of enquiry was applied to the description and analysis of consciousness through which philosophy attempts to gain the character of a strict science. Husserl's method is an effort to resolve the opposition between the emphasis on observation that is maintained by Empiricism, and on reason that is stressed by Rationalism.

Against this background, Sartre maintained that Existentialism is an attempt to live logically in a universe that is ultimately absurd. Another eloquent supporter of this doctrine was the Literature Nobel Laureate Albert Camus. In the mid-20th century Camus, through writings addressed the isolation of man in what he considered to be an alien universe[14]. At the end of this long line of intellectuals that were under the influence of Existentialism, in the above quotation Weinberg reflects a view of the universe to which he was constrained by the adopted philosophic trend that influenced his generation.

In the existentialist view of the universe there still remains some hope for the concept of a meaningful universe, namely intelligibility of the universe could be approached with the hope of the eventual emergence of a future "theory of everything". In this proposed all-embracing future theory we would hopefully discover the fundamental laws of nature in terms of a set of equations[15]. Then, all phenomena should follow from these equations (the hope being that chemistry and biology could also be deduced). This is an extreme form of reductionism, not an inevitable choice, given the many insights of current progress in science as a whole. Since the Enlightenment the ever-increasing growth of science has encouraged reductionism. The ideas that physical bodies are collections of atoms or that thoughts are combinations of sense impressions are forms of reductionism. Philosophers have held two very general forms of reductionism in the 20th century.

Firstly, logical positivists have maintained that expressions referring to existing things or to states of affairs are definable in terms of directly observable objects, or sense-data, and, hence, that any statement of fact is equivalent to some set of empirically verifiable statements. In particular, it has been held that the theoretical entities of science are definable in terms of observable physical things, so that scientific laws are equivalent to combinations of observation reports.

Secondly, proponents of the unity of science have held the position that the theoretical entities of particular sciences, such as biology or psychology, are definable in terms of more basic science, such as physics. In other words, laws of these sciences can be explained in terms of more basic science. The logical positivist version of reductionism also implies the unity of science insofar as the definability of the theoretical entities of the various sciences in terms of the observable would constitute the common basis of all scientific laws. Although this version of reductionism is no longer widely accepted, primarily because of the difficulty of giving a satisfactory characterization of the distinction between theoretical and observational statements in science, the question of the reducibility of one science to another remains controversial.

The reductionist dream has been supported by preliminary sets of successful equations that have embodied general phenomena at the most disparate scales (both microscopic and macroscopic). Today we recognize such efforts by assigning the equations the surnames of their authors: Newton, Einstein, Dirac, Salam, Schrödinger and Weinberg. We are still at a very early stage in the comprehension of life in the universe. When the open question of the intelligibility of the universe is posed in a wider cultural context, including the earth and life sciences, Reductionism's restricted view becomes more evident. This contrast between different views on the intelligibility of the universe illustrates the relevance of the frontier of astrobiology for human culture, especially under the influence of philosophical doctrines (other than Existentialism) that will tend to encourage any future constructive dialogue between science and the humanities. In the remaining part of this chapter we shall attempt to review briefly what science has achieved in our understanding of the destinies. We begin with the destiny of the universe.

14.2 What is the likely destiny of life?

Having reviewed our present understanding of the eventual destiny of the cosmos, our next objective, as stated in the title chosen for this chapter, is to discuss the destiny of the phenomenon of the living process that has emerged in the universe. The origin of life is not fully understood. However, the general outline of the question of the chemical evolution of the precursors of the biomolecules has greatly advanced. Likely pathways have been outlined that nature may have followed during the molecular evolution that preceded the Darwinian evolution of the living cell. This topic has been reviewed in Chapters 3. The seminal work of Darwin established the basis for the second stage in the discussion of astrobiology. Darwinian evolution is much better understood than the question of the emergence of life.

In Chapters 7 and 8 we have argued that evolution on Earth has taught us that evolutionary convergence is an important feature of the Earth biota. Hence, if Darwinian evolution were assumed to be a universal process[16], we would expect that whenever life emerges elsewhere in the universe, life would be bound largely by the same general properties that we have found on Earth. Under this assumption (the universality of biology), we can anticipate new insights in the distribution of life in the universe.

Perhaps the leading approaches for the search of life elsewhere in the cosmos are the exploration of the Solar System and the search for intelligent signals through windows of the electromagnetic spectrum. The latter—the SETI project-was already introduced in Sec. 2.1. Since the pioneering days of the 1960s, bioastronomers have probed various windows of the electromagnetic spectrum for evidence of narrow bandwidth signals[17,18]. Such output presumably would be characteristic of other civilizations, instead of being the product of natural phenomena, such as supernova explosions or regular emissions from pulsars. The search for other intelligent civilizations in the SETI project might have some implications in the philosophy of religion.

Our religious traditions go back to Jewish theology. There is a sole omnipotent God who created heaven and earth, and subsequently life on earth. This view of our origins has traditionally been referred to as a "first" Genesis. But revelation through the scriptures never raises explicitly the possibility of the plurality of inhabited worlds.

In 1584 Giordano Bruno made a speculative, but significant reference to the possibility of ubiquitous life in the universe[19] (cf., Sec. 5.5). In the late 16th century Bruno's statement led to a bitter and tragic controversy in the frontier between science and religion. However, due to the present progress both in science and religion, we are now aware that there is no evident incompatibility between religious traditions and the possibility that we may not be alone in the universe. What is exciting about the emergence of the new science of astrobiology is that we can explore in strictly scientific terms the possibility of whether the evolution of intelligent behavior is inevitable in an evolving cosmos, as already assumed implicitly by the above-mentioned SETI project.

14.3 The destiny of life and a second Genesis

In the previous two sections we have looked firstly at the possible destiny of the universe. Secondly, we considered the living process that has emerged through chemical and biological evolution on Earth, and possibly elsewhere. The next step in our discussion is dictated by the fact that the theory of Darwinian evolution is not a predictive theory. In order to get further insights as to what can be the eventual destiny of life in the universe that might not be evident from our knowledge of biology, we should search for alternative manifestations life.

This may occur either in planets, or satellites in our own solar system, or even in other solar systems. Intelligent signals from other civilizations are in principle detectable. The fact remains though that our lives are short we would like to have further insights into our destiny.

After almost half a century of searching for intelligent life in the universe—with extraordinary technological progress in the detection equipment used in the SETI project—sadly, no intelligent signals have so far been identified. But technology not only has progressed in recent years in the field of bioastronomy, it has also progressed especially in the exploration of the Solar System with missions planned by the main space agencies, so as to be in principle capable of detecting microscopic life.

The search for extraterrestrial life was attempted for the first time on the surface of planet Mars a quarter of a century ago: the Viking missions were in a position to detect life, although their results were not convincing to most scientists. The search continues today with Mars being the present target of several space missions that have been given ample publicity in the media.

Yet, given the harsh conditions for the survival of extremophilic microorganisms on the Red Planet, the best digging equipment with present technology is still unable to probe as far as the more likely sites, deep underground, where we expect abundant liquid water to be present. Having argued in Chapter 9 that Darwinian evolution is a universal process, we may be confronted with the inevitability of the origin and evolution of life, including intelligence. We have also argued and that the role of contingency has to be seen in the restricted context of evolutionary convergence[20,21].

14.4 Further thoughts on the universality of science

Convergence is not restricted to biology, but it has some relevance in other realms of science. The sharp distinction between chance (contingency) and necessity (natural selection as the main driving force in evolution) is relevant for astrobiology. For this reason, it is important to document the phenomenon of evolutionary convergence at all levels, in the ascent from stardust to brain evolution. In particular, documenting evolutionary convergence at the molecular level is the first step in this direction (cf., Sec. 8.4).

The universality of biochemistry suggests that in Solar System missions, biomarkers should be selected from standard biochemistry. Given the importance of deciding whether the evolution of intelligent behavior has followed a convergent evolutionary pathway, and given the intrinsic difficulty of testing these ideas directly, we can alternatively begin testing the lowest stages of the evolutionary pathway within the Solar System. Within a few years we will be in a position to search directly for evolutionary biomarkers on Europa, the Jovian satellite.

We have considered that if extant microorganisms were to be encountered, a possible set of evolutionary biomarkers may be considered[22]. In addition, if only landers are possible in the foreseeable future, it is possible with the help of available techniques in biogeochemistry to decide on the possible biomarkers by sampling regions of the icy surface of Europa where there is abundant presence of non-ice elements such as, for example, sulfur[23].

Testing evolutionary biomarkers clearly lies in the distant future. However, we expect that technology will allow us to be in a position to undertake a variety of experiments *in* situ on the surface of other solar system bodies, such as the Jovian moon Europa. Given the length of time before we can test reliable biomarkers directly, a full discussion at the present time of the feasibility of carrying out a proper test is timely. In this sense the discussion of biomarkers is reasonable at the present stage, since the Galileo mission has already provided us with a wealth of information about the chemical non-water-ice elements on the icy surface[24,25]. The careful interpretation of such information might conceivably lead us to reliable biomarkers without actually penetrating Europa's icy surface. Testing directly the Europan icy surface with a lander is a possibility that can be taken into account.

The discovery of other solar systems suggests that their formation seem to be analogous to ours. This is compatible with extensive knowledge of interstellar matter[26]. From the assumed universality of biology it seems inevitable that intelligent behavior will emerge in the cosmos, provided certain conditions favorable to the presence of continuous life on a given planet (or satellite) are maintained. One of these conditions is that early stages in the formation of a solar system are characterized by a heavy bombardment period. This period would end after a few hundred million years. Consequently, planetary conditions over geologic time are likely to allow the continuous presence of life, as it has already occurred on our own planet, once the heavy bombardment ceased. Observational techniques continually improve (the Terrestrial Planet Finder and the large telescopes of the European Space Observatory are now under construction). These new instruments will allow us to estimate the duration of the initial heavy bombardment in other solar systems, as well as the subsequent life-favorable quiescent period. These future observations should give us a more precise idea of the temporal constraints that allow the continuous presence of life on a given planet. In our own solar system the most attractive site for the search for life has already been explored.

As we mentioned in Sec. 9.5, the Galileo mission to Jupiter and the Galilean satellites completed an eight-year period of continuous exploration in the year 2003. This mission focused our attention on Europa. In the 17th century Galileo discovered Europa together with three other satellites Io, Ganymede and Callisto. However, Europa remains the leading contender for being the host of an independent evolutionary line. A second Genesis could, in principle, be brought to our attention in the foreseeable future if the funding of the space agencies allows this technical possibility.

Ganymede is the largest satellite in the Solar System, bigger than the planets Mercury and Pluto). In fact, the Galileo mission already has given us data to suggest that Ganymede, Callisto and Europa may harbor large oceans underneath their icy surfaces. Not only is there strong evidence for the internal oceans in the Jovian system, but also Jupiter's large icy moons appear to have three ingredients essential for the origin and sustained evolution of life, namely, water, energy and the necessary chemical molecules.

It is still premature to attempt the exploration of extraterrestrial oceans of the Jovian system by means of submersibles (cf., Sec. 9.5). But for learning whether a second Genesis has occurred, probably a lander may be sufficient. We should recall some aspects of the icy Europan surface: it is possible that matter from the interior of the satellite may be raised to the surface itself. The Galileo mission has led to the discovery of a phenomenon called "lenticulae" that are interpreted as surface areas on Europa, whose origin is matter from its deep interior. A lander, without penetrating the icy shell, may be sufficient to retrieve information that might shed some light on its subsurface ocean that may be pregnant with life.

14.5 A dialogue across the frontiers of science and the humanities

Throughout the book we have considered a common frontier of astrobiology, philosophy, as well as other branches of the humanities. We have attempted to show that an interdisciplinary exchange across the border is not only possible, but also it may be profitable for the whole of human culture.

A conflict is not expected to arise with the potential discovery of a second Genesis. Instead, a dialogue could emerge with a discussion of the evolution of all the attributes of man, including those that are of prime importance for theology, namely the human spirit that may distinguish us from the ancestors of the *Homo* genus.

However, if we remain strictly within the constraints that the science of biology has imposed onto itself—namely that the life sciences are based on observation (for instance, the evidence supporting natural selection), or on experiment (for example, in biochemistry), the question of man's spirit and soul should not enter into biology, but a dialogue across the frontier of science should always be encouraged.

Bibliography

Preface

1. Russell, B. (1991) *History of Western Philosophy and Its Connection with Political and Social Circumstances from the Earliest Times to the Present Day*, Routledge, London, p. 13.

2. Walker, C. B. F. (1987) *Reading the Past Cuneiform*, British Museum, London, p. 9.

3. Watterson, B. (1981) *Introducing Egyptian Hieroglyphs*, Scottish Academic Press, Edinburgh, p. 6.

4. Cornwell, J. (2007) *Darwin's Angel*: *A Seraphic Response to "The God Delusion"*, Profile Books, London.

Acknowledgments

1. Salam, A. (1990) *Ideals and Reality*, 3rd edition. C. H. Lai (ed.), World Scientific, Singapore.

2. Salam, A. (1991) The role of chirality in the origin of life. *J. Mol. Evol.* **33**, 105-113.
   This publication has been reproduced in the highly recommended collection of Abdus Salam's scientific papers (cf., Re. 3).

3. Ali, A., Isham, C., Kibble, T. and Riazuddin (1994) *Selected Papers of Abdus Salam with Commentary*, World Scientific, Singapore, pp. 626-634. (cf., Appendix D where Salam agrees that Nobel Laureate Max Delbruck had anticipated the significance of condensation in biology, a seminal remark that inspired Salam's work.)

4. By special request of Salam the following paper was written to explain his theory to a wider public: Chela-Flores, J. (1991) Comments on a novel approach to the role of chirality in the origin of life. *Chirality* **3**, 389-392.

5. Chela-Flores, J. (1995) Is the Salam phase transition relevant to the causal origin of homochirality? *Proc. Pakistan Acad. Sci.* **32**, 1-12. (By invitation of the Editor, Dr. Hameed Ahmed Khan, SI in the dedication of a special issue of "The Nucleus" to Professor Dr. Abdus Salam).

6. Ponnamperuma, C. and Chela-Flores, J. (eds.) (1993) *Chemical Evolution*: *Origin of Life*, Vol. **135**, A. Deepak Publishing, Hampton, Virginia, USA.

7. Ponnamperuma, C. and Chela-Flores, J. (Guest Editors) (1995) Chemical evolution: The structure and model of the first cell, *J. Biol. Phys.* **120** 1-4.

8. Chela-Flores, J., Lemarchand, G. A. and Oro, J. (2000) *Astrobiology*: *Origins from the Big Bang to Civilization*, Kluwer Academic Publishers, Dordrecht, The Netherlands.

9. Chela-Flores, J, Owen, T. and Raulin, F. (eds.) (2001) *The First Steps of Life in the Universe*, Kluwer Academic Publishers, Dordrecht, The Netherlands.

10. Seckbach, J., Chela-Flores, J., Owen, T. and Raulin, F. (eds.) (2004*) Life in the Universe from the Miller Experiment to the Search for Life on Other Worlds*, Series: Cellular Origin, Life in Extreme Habitats and Astrobiology, Vol. **7**, Springer, Dordrecht.

11. Chela-Flores, J. and Raulin, F. (eds.) (1996) *Chemical Evolution: Physics of the Origin and Evolution of Life* (The Cyril Ponnamperuma Memorial Conference), Kluwer Academic Publishers, Dordrecht, The Netherlands.

12. Chela-Flores, J. and Raulin, F. (eds.) (1998) *Exobiology: Matter, Energy, and Information in the Origin and Evolution of Life in the Universe,* Kluwer Academic Publishers, Dordrecht, The Netherlands.

13. Mayz-Vallenilla, E. (2003) Astrofísica y Meta-técnica, *Letras de Deusto*, Universidad de Deusto, Bilbao, Spain, Vol. XXXIII, No. 98, January-March 2003, pp. 217-224.

14. Chela-Flores, J. (2003) Marco cultural de la Astrobiología, *Letras de Deusto*, Universidad de Deusto, Bilbao, Spain, Vol. XXXIII, No. 98, January-March 2003, pp. 199-215.

15. Aretxaga, R. and Chela-Flores, J. (2006) Astrobiologia y Filosofia (II), *Letras de Deusto*, Universidad de Deusto, Bilbao, Vol. XXXVI, No. 110, pp. 9-36.

16. Aretxaga, R. (2003) Astrobiología y Filosofía. Presentación. *Letras de Deusto*, Universidad de Deusto, Bilbao, Spain, Vol. XXXIII, No. 98, January-March 2003, pp. 189-198.

17. Aretxaga Burgos, R. (2003) La ciencia astrobiologica. Un nuevo reto pare el humanismo del siglo XXI, Congreso Internacional: Humanismo para el Siglo XXI, Universidad de Deusto, Bilbao, Spain, 4-7 March, 2003.

18. Aretxaga Burgos, R. (2003) Astrobiology and Biocentrism, in *Life in the Universe*, J. Seckbach, J. Chela-Flores, T. Owen and F. Raulin (eds.), Series: Cellular Origin and Life in Extreme Habitats and Astrobiology, Vol. **7**, Springer, Dordrecht, The Netherlands, pp. 345-348.

19. Chela-Flores, J. (1997) Cosmological models and appearance of intelligent life on Earth: The phenomenon of the eukaryotic cell, in *Reflections on the Birth of the Universe: Science, Philosophy and Theology*, P. Eligio, G. Giorello, G. Rigamonti and E. Sindoni (eds.), Edizioni New Press, Como, pp. 337-373.

20. Chela-Flores, J. (1998) The phenomenon of the eukaryotic cell, in *Evolutionary and Molecular Biology: Scientific Perspectives on Divine Action*, R. J. Russell, W. R. Stoeger and F. J. Ayala (eds.). Vatican City State/Berkeley, California, Vatican Observatory and the Center for Theology and the Natural Sciences, pp. 79-99.

21. Chela-Flores, J. (1999) Gli alberi della vita, in Carlo Maria Martini *Orizzonti e limiti della scienza, Decima Cattedra di non credenti*. E. Sindoni and C. Sinigaglia (eds.). (In the series "Scienze e Idee", directed by G. Giorello.) Raffaello Cortina Editore, Milano. pp. 43-50.

22. Martini, C. M. and Chela-Flores, J. (1999) Dialogo, in Carlo Maria Martini *Orizzonti a limiti della scienza*, *Decima Cattedra di non credenti*, E. Sindoni and C. Sinigaglia (eds.). (In the series "Scienze e Idee", directed by G. Giorello.) Raffaello Cortina Editore, Milano, pp. 65-68.

23. Martini, C. M. (2006*) Il mio novecento*, Centro Ambrosiano, Milano, pp. 64-66.

24. Chela-Flores, J. (2008) Fitness of the cosmos for the origin and evolution of life: From biochemical fine-tuning to the Anthropic Principle, in *Fitness of the Cosmos for Life*: *Biochemistry and Fine-tuning*, J. D. Barrow, S. Conway Morris, S. J. Freeland and C. L. Harper (eds.), Cambridge University Press, pp. 151-166.

25. Coyne, S. J., G. (1996) Cosmology: The Universe in Evolution, in *Chemical Evolution: Physics of the Origin and Evolution of Life*, J. Chela-Flores and F. Raulin (eds.), (The Cyril Ponnamperuma Memorial Conference), Kluwer Academic Publishers, Dordrecht, The Netherlands, pp. 35-49.

26. Coyne, S. J., G. (1998) The concept of matter and materialism in the origin and evolution of life, in *Exobiology: Matter, Energy, and Information in the Origin and Evolution of Life in the Universe*, J. Chela-Flores and F. Raulin (eds.), Kluwer Academic Publishers, Dordrecht, The Netherlands, pp. 53-60.

27. Coyne, S. J., G. (2001) Origins and creation, in *The First Steps of Life in the Universe*, J. Chela-Flores, T. Owen and F. Raulin (eds.), Kluwer Academic Publishers, Dordrecht, The Netherlands, pp. 359-364.

28. Coyne, S. J., G. (2004) An overview of cosmic evolution, in *Life in the Universe from the Miller Experiment to the Search for Life on Other Worlds*, J. Seckbach, J. Chela-Flores, T. Owen and F. Raulin (eds.), Series: Cellular Origin, Life in Extreme Habitats and Astrobiology, Vol. **7**, Springer, Dordrecht, The Netherlands, pp. 17-26.

29. Drake, F. (1996) Progress in searches for extraterrestrial intelligent radio signals, in *Chemical Evolution*: *Physics of the Origin and Evolution of Life*, J. Chela-Flores and F. Raulin (eds.), (The Cyril Ponnamperuma Memorial Conference), Kluwer Academic Publishers, Dordrecht, The Netherlands, pp. 335-341.

30. Drake, F. (1998) The Cyril Ponnamperuma Lecture: The search for intelligent life in the universe in *Exobiology: Matter, Energy, and Information in the Origin and Evolution of Life in the Universe*, J. Chela-Flores and F. Raulin (eds.), Kluwer Academic Publishers, Dordrecht, The Netherlands, pp. 71-80.

31. Drake, F. (2001) Pre-dinner speaker: New paradigms for SETI, in *The First Steps of Life in the Universe*, J. Chela-Flores, T. Owen and F. Raulin (eds.), Kluwer Academic Publishers, Dordrecht, The Netherlands, pp. 393-398.

Chapter 1

1. Haught, J. F. (2005) Darwin, design and promise of nature (The Boyle Lecture), *Science Christian Belief*, **17** (1), 5-20.

2. Sagan, C. (1995) *Pale Blue Dot a Vision of the Human Future in Space*, Headline Book Publishing, London.

3. Chela-Flores, J. (2001) *The New Science of Astrobiology from Genesis of the Living Cell to Evolution of Intelligent Behavior in the Universe*, Kluwer Academic Publishers, Dordrecht, The Netherlands.

4. Marino, L. (1998) A comparison of encephalization between odontocete cetaceans and anthropoid primates, *Brain Behavior, Evol.* **51**, 230-238.

5. Dante A. (1985) *La Divina Commedia*, Oscar Mondadori, Milano, Paradiso, Canto XVIII, 95-96 (...si' che Giove pareva argento li' d'oro distinto).

6. Greenberg, R. (2005) *Europa the Ocean Moon*, Springer-Praxis Books in Geophysical Sciences, Chichester, UK, Figure 2.2.

7. Horvath, J., Carsey, F., Cutts, J., Jones, J., Johnson, E., Landry, B., Lane, L., Lynch, G., Chela-Flores, J., Jeng, T-W. and Bradley, A. (1997) Searching for ice and ocean biogenic activity on Europa and Earth, in *Instruments, Methods and Missions for Investigation of Extraterrestrial Microorganisms*, R. B. Hoover (ed.), Proc. SPIE, Vol. **3111**, pp. 490-500.

8. Chela-Flores, J. (2006) The sulphur dilemma: Are there biosignatures on Europa's icy and patchy surface? *Int. J. Astrobiol.* **5**, 17-22 (Cambridge University Press).

9. Smith, A., Crawford, I. A., Gowen, R. A., Ball, A. J., Barber, S. J., Church, P., Coates, A. J., Gao, Y., Griffiths, A. D., Hagermann, A., Phipps, A., Pike, W. T., Scott, R., Sheridan, S., Sweeting, M., Talboys, D., Tong, V., Wells, N., Biele, J., Chela-Flores, J., Dabrowski, B., Flannagan, J., Grande, M., Grygorczuk, J., Kargl, G., Khavroshkin, O. B., Klingelhoefer, G., Knapmeyer, M., Marczewski, W., McKenna-Lawlor, S., Richter, L., Rothery, D. A., Seweryn, K., Ulamec, S., Wawrzaszek, R., Wieczorek, M. and Wright, I. P. (2008) LunarEX—A proposal to Cosmic Vision, *Exp. Astron.* 10.1007/s10686-008-9109-6 (August 21, 2008).

10. Blanc, M., and the LAPLACE consortium (2008) LAPLACE: A mission to Europa and the Jupiter system, *Astrophys. Instrum. Meth.* in press.

Chapter 2

1. Henderson, L. J. (1913) *The Fitness of the Environment: An Enquiry into the Biological Significance of the Properties of Matter*, Peter Smith, Gloucester, Mass., 1970.

2. Chela-Flores, J. (2006) Destinies of life and the universe: the final frontiers of astrobiology and cosmology, in *"Life as we Know It"*. Cellular Origins, Life in Extreme Habitats and Astrobiology Springer, Dordrecht, The Netherlands, pp. 505-517.

3. Barrow, J. D., Conway Morris, S., Freeland, S. J. and Harper, C. L. (eds.) (2006) *Fitness of the Cosmos for Life: Biochemistry and Fine-Tuning*, Cambridge University Press.

4. Ekers, R. D., Kent Cullers, D., Billingham, J. and Scheffer, L. K. (eds.) (2002) *SETI 2020*. SETI Press, Mountain View, CA.

5. Baines, J. and Malek, J. (1980) *Atlas of Ancient Egypt*, Phaidon, Oxford.

6. Nepi Scire, Giovanna, (ed.) (1998) *The Accademia Galleries in Venice*, Electa, Milan.

7. Armstrong, Karen (1997) *Jerusalem One City Three Faiths*, Ballantine Books, New York.

8. Chapman, Allan (2001) *Gods in the Sky Astronomy from the Ancients to the Renaissance*, Channel 4 Books, London.

9. Russell, B. (1991) *History of Western Philosophy and Its Connection with Political and Social Circumstances from the Earliest Times to the Present Day*, Routledge, London, pp. 44-48.

10. Plato (1974) *The Republic*, Penguin Classics, London.

11. Aristotle (1998) *Nicomachean Ethics*, Dover, New York.

12. Williams, B. (1982) Evolution and ethics, in *Evolution From Molecules to Men*, D. S. Bendall (ed.), Cambridge University Press, London, pp. 555-566.

13. Conway-Morris, S. (1998) *The Crucible of Creation*, Oxford University Press.

14. Conway-Morris, S. (2003*) Life's Solution Inevitable Humans in a Lonely Universe*, Cambridge University Press.

15. Akindahunsi, A. A. and Chela-Flores, J. (2004) On the question of convergent evolution in biochemistry, in *Life in the Universe: From the Miller Experiment to the Search for Life on Other Worlds*, J. Seckbach, J. Chela-Flores, T. Owen and F. Raulin (eds.), Kluwer Academic Publishers, Dordrecht, pp. 135-138.

16. Enlightenment is also known as the age of reason. During this period there was a remarkable development of an empiricist, and materialism was emphasized in opposition to clericalism, notably in the publication of the Encyclopedie. This work was published by a group of French writers, especially Diderot and D'Alembert.

17. Newton, I. (1687) *De Philosophiae Naturalis Principia Mathematica*, London.

18. Voltaire (pseudonym of François-Marie Arouet) (1764) *Dictionnaire Philosophique,* English translation by H. I. Woolf, 1945.

19. Feigel, H. (1963) Physicalism, unity of science and the foundations of psychology, in *The Philosophy of Rudolf Carnap*, P. A. Schilpp (ed.) (The Library of Living Philosophers, Vol. **11**), Cambridge University Press, London, pp. 227-267.

20. Russell, B. (1991) *History of Western Philosophy and Its Connection with Political and Social Circumstances From the Earliest Times to the Present Day*, Routledge, London.

21. John Paul II (1992) Discorso di Giovanni Paolo II alla Pontificia Accademia delle Scienze. *L'Osservatore Romano*, 1st November, p. 8.

22. Chela-Flores, J. (1997) Cosmological models and appearance of intelligent life on Earth: The phenomenon of the eukaryotic cell, in *Reflections on the Birth of the Universe: Science, Philosophy and Theology*, P. Eligio, G. Giorello, G. Rigamonti and E. Sindoni (eds.), Edizioni New Press, Como, pp. 337-373.

23. Wilson, E. O. (1998) *Consilience The Unity of Knowledge*, Alfred A. Knopf, New York, pp. 3-7.

24. Schilpp, P. A. (ed.) (1963) *The Philosophy of Rudolf Carnap*. The Library of Living Philosophers, Vol. **11**, London, Cambridge University Press. In pp. 227-267 Herbert Feigl summarizes the conclusions arrived during these discussions.

25. Simpson, W. K. (ed.) (1972) *The Literature of Ancient Egypt. An Anthology of Stories, Instructions and Poetry*, Yale University Press, New Haven, pp. 7-9, 289-295.

26. Russell, B. (1991) *History of Western Philosophy and Its Connection with Political and Social Circumstances From the Earliest Times to the Present Day*, Routledge, London, p. 542.

27. Drake, F. and Sobel, D. (1992) *Is There Anyone Out There? The Scientific Search for Extraterrestrial Intelligence*, Delacorte Press, New York, pp. 45-64.

28. Gould, S. J. (1991) *Wonderful life. The Burgess Shale and the Nature of History*, Penguin Books, London, pp. 48-52.

29. Conway-Morris, S. (1998) *The Crucible of Creation*, Oxford University Press, pp. 12-14.

30. Tucker Abbott, R. (1989) *Compendium of Landshells*, American Malacologists, Melbourne, Florida, USA, pp. 7-8.

31. Mayor, M., Queloz, D., Udry, S. and Halbwachs, J.-L. (1997) From brown dwarfs to planets, in *Astronomical and Biochemical Origins and the Search for Life in the Universe*, C. B. Cosmovici, S. Bowyer and D. Werthimer (eds.), Editrice Compositore, Bologna, pp. 313-330.

32. Teilhard de Chardin, P. (1965) *The Phenomenon of Man*, Fontana Books, London, p. 33.

33. Chela-Flores, J. (1998) The phenomenon of the eukaryotic cell, in *Evolutionary and Molecular Biology: Scientific Perspectives on Divine Action*, R. J. Russell, W. R. Stoeger and F. J. Ayala (eds.), Vatican City State/Berkeley, California, Vatican Observatory and the Center for Theology and the Natural Sciences, pp. 79-99.

34. De Duve, C. (1995) *Vital Dust: Life as a Cosmic Imperative*, Basic Books, New York, pp. 160-168.

35. John Paul II (1996) Papal Message to the Pontifical Academy, *Commentarii* 4(3) Vatican City, 1997 (pp. 15-20), cf., also: "La traduzione in italiano del Messaggio del Santo Padre alla Pontificia Accademia delle Scienze", L'Osservatore Romano, 24 October 1996, p. 7.

36. Dick, S. J. (2004) The new universe, destiny of life, and cultural implications, in *Life in the Universe: From the Miller Experiment to the Search for Life on Other Worlds*, in J. Seckbach, J. Chela-Flores, T. Owen and F. Raulin (eds.), Kluwer Academic Publishers, Dordrecht, pp. 319-326. The reader is also advised to refer to Dick, S. J. (1998) *Life on Other Worlds*, Cambridge University Press, London, pp. 245-256.

37. Ward, P. D. and Brownlee, D. (2000) *Rare Earth. Why Life is Uncommon in the Universe*, Copernicus, Springer-Verlag, New York.

38. Hoyle, F. (1975) *Astronomy and Cosmology: A Modern Course*, W. H. Freeman and Company, San Francisco, pp. 401-402.

39. Dunbar, D. N. F., Pixley, R. E., Wenzel, W. A. and Whaling, W. (1953) The 7.68-Mev state in C^{12}, *Phys. Rev.* **92**, 649-650.

40. Ponnamperuma, C. and Chela-Flores, J. (eds.) (1993) *Chemical Evolution: Origin of Life*, A. Deepak Publishing, Hampton, Virginia, USA.

41. Chela-Flores, J., Owen, T. and Raulin, F. (eds.) (2001) *The First Steps of Life in the Universe*, Kluwer Academic Publishers, Dordrecht, The Netherlands.

42. Chela-Flores, J. (2008) Fitness of the cosmos for the origin and evolution of life: From biochemical fine-tuning to the Anthropic Principle, in *Fitness of the Cosmos for Life: Biochemistry and Fine-Tuning*, J. D. Barrow, S. Conway Morris, S. J. Freeland and C. L. Harper (eds.), Cambridge University Press, pp.151-166.

43. Haught, J. F. (1998) Darwin's gift to theology, in *Evolutionary and Molecular Biology: Scientific Perspectives on Divine Action*, R. J. Russell, W. R. Stoeger S. J. and F. J. Ayala, (eds.), Vatican City State/Berkeley, California, Vatican Observatory and CTNS, pp. 393-418.

44. Polkinghorne, J. (1996) *Scientists as Theologians*, SPCK, London.

45. Rovelli, C. (2000) The century of the incomplete revolution: searching for a general relativistic quantum field theory, *J. Math. Phys.* **41**, 3776-800.

46. Krubitzer, L. (1995) The organization of neocortex in mammals: are species differences really so different? *Trends Neurosci.* **18**, 408-417.

47. Monod, J. (1972) *Chance and Necessity an Essay on the Natural Philosophy of Modern Biology*, Collins, London.

48. Peacocke, A. (1988) Biological evolution—a positive theological appraisal, in *Evolutionary and Molecular Biology: Scientific Perspectives on Divine Action*. R. J. Russell, W. R. Stoeger and F. J. Ayala (eds.), Vatican City State/Berkeley, California, Vatican Observatory and the Center for Theology and the Natural Sciences, pp. 357-376.

49. Some of the intriguing possibilities of a first contact between different civilizations have been presented in fictional form. As the technological possibilities for such an achievement in the future are well documented, the reader is recommended to read: Sagan, C. (1985) *Contact*, Simon & Schuster, New York.

Chapter 3

1. Flew, A. (ed.) (1979) *A Dictionary of Philosophy*, Pan Books, London, pp. 300-301.

2. Henderson, L. J. (1913) *The Fitness of the Environment An Enquiry into the Biological Significance of the Properties of Matter*, Peter Smith, Gloucester, Mass., 1970, p. 312.

3. Holmes, S. J. (1961) *Louis Pasteur*, Dover, New York, p. 51.

4. Russell, R. J. (2001) Life in the universe: Philosophical and theological issues, in *The First Steps of Life in the Universe*, J. Chela-Flores, T. Owen and F. Raulin (eds.), Proceedings of the Sixth Trieste Conference on Chemical Evolution. Trieste, Italy, 18-22 September, Kluwer Academic Publishers, Dordrecht, The Netherlands, pp. 365-374.

5. Ayala, F. J. (1998) Darwin's devolution: Design without designer, in *Evolutionary and Molecular Biology: Scientific Perspectives on Divine Action*, R. J. Russell, W. R Stoeger S. J. and F. J. Ayala, (eds.), Vatican City State/Berkeley, California, Vatican Observatory and the Center for Theology and the Natural Sciences, pp. 101-116.

6. Monod, J. (1972) *Chance and Necessity an Essay on the Natural Philosophy of Modern Biology,* Collins, London.

7. Snow, C. P. (1978) *The Two Cultures and a Second Look*, Cambridge University Press.

8. St. Augustine (1984) *Concerning the City of God Against the Pagans*, Penguin Classics, London (cf., Book XVII, p. 4).

9. John Paul II (1996) Papal Message to the Pontifical Academy of Sciences of 22 October 1996, *L'Osservatore Romano Weekly Edition.* No. 44, 30 October, p. 3, 7. (A translation from the official version in the French Language.)

10. Polkinghorne, J. (1996) *Scientists as Theologians*, SPCK, London.

Chapter 4

1. This chapter is a development of earlier papers in Chela-Flores, J. (2005) Fitness of the universe for a second Genesis: Is it compatible with science and christianity? *Science and Christian Belief* **17** (2), pp. 187-197 and in Chela-Flores, J. (2007) Fitness of the cosmos for the origin and evolution of life: From biochemical fine-tuning to the Anthropic Principle, in *Fitness of the Cosmos for Life: Biochemistry and Fine-Tuning*, J. D. Barrow, S. Conway Morris, S. J. Freeland and C. L. Harper, (eds.), Cambridge University Press, pp. 151-166.

2. Russell, B. (1995) *Religion and Science,* Dover, New York.

3. McKay, C. P. (2001) The search for a second Genesis in our Solar System, in *The First Steps of Life in the Universe*, J. Chela-Flores, T. Owen and F. Raulin (eds.), Kluwer Academic Publishers, Dordrecht, pp. 269-277.

4. De Duve, C. (1995) *Vital Dust. Life as a Cosmic Imperative*, Basic Books, A Division of Harper-Collins Publishers, New York, pp. 296-297.

5. Crick, F. (1981) *Life Itself Its Origin and Nature*, Macdonald, London.

6. Ward, P. D. and Brownlee, D. (2000) *Rare Earth: Why Complex Life is Uncommon in the Universe*, Copernicus, New York.

7. Dawkins, R. (1983) Universal darwinism, in *Evolution From Molecules to Men*, D. S. Bendall, (ed.), Cambridge University Press, London, pp. 403-425.

8. Akindahunsi, A. A. and Chela-Flores, J. (2004) On the question of convergent evolution in biochemistry, in *Life in the Universe: From the Miller Experiment to the Search for Life on Other Worlds*, J. Seckbach, J. Chela-Flores, T. Owen and F. Raulin (eds.), Kluwer Academic Publishers, Dordrecht, pp. 135-138.

9. Fontana, W. and Buss, L. W. (1994) What would be conserved if "the tape were played twice?", *Proc. Natl. Acad. Sci. USA* **91**, 757-761.

10. Chela-Flores, J. (2001) *The New Science of Astrobiology From Genesis of the Living Cell to Evolution of Intelligent Behaviour in the Universe*, Kluwer Academic Publishers, Dordrecht (paperback edition, 2004), http://www.wkap.nl/prod/b/0-7923-7125-9.

11. Chela-Flores, J. (2004) Astrobiology's last frontiers: The distribution and destiny of life in the universe, in *Origins: Genesis, Evolution and the Biodiversity of Life*, J. Seckbach (ed.), Kluwer Academic Publishers, Dordrecht, The Netherlands, pp. 667-679.

12. Conway Morris, S. (1998) *The Crucible of Creation The Burgess Shale and the Rise of Animals*, Oxford University Press, London, p. 202.

13. Conway Morris, S. (2003) *Life's Solution Inevitable Humans in a Lonely Universe*, Cambridge University Press, London.

14. Drake, F. and Sobel, D. (1992) *Is There Anyone Out There? The Scientific Search for Extraterrestrial Intelligence*, D. Delacorte Press, New York.

15. Ekers, R. D., Kent Cullers, D., Billingham, J. and Scheffer, L. K., (eds.) (2002) *SETI 2020*, SETI Press, Mountain View California, USA.

16. Chela-Flores, J. (2003) Testing evolutionary convergence on Europa, *Int. J. Astrobiol.* **2** (4), 307-312, Cambridge University Press.

17. Johnson, T. (2004) Europa: overview of the future missions, in *Life in the Universe From the Miller Experiment to the Search for Life on Other Worlds*, J. Seckbach, J. Chela-Flores, T. Owen and F. Raulin (eds.), Series: Cellular Origin, Life in Extreme Habitats and Astrobiology, Vol. **7**, pp. 3-5.

18. John Paul II (1997) Papal Message to the Pontifical Academy, *Commentarii*, Vatican City **4** (3), 15-20.

19. Coyne S. J., G. (1998) The concept of matter and materialism in the origin and evolution of life, in *Chemical Evolution: Exobiology. Matter, Energy, and Information in the Origin and Evolution of Life in the Universe*, J. Chela-Flores and F. Raulin (eds.), Kluwer Academic Publishers, Dordrecht, pp. 71-80.

20. Dewey, J. (1951) *The Influence of Darwinism on Philosophy*, Peter Smith, New York.

21. Maritain, J. (1960) *La philosophie morale. I. Examen historique et critique des grands systemes*, Gallimard, Bibliothèque de Idées, Paris.

22. Russell, R. (2001) Life in the universe: Philosophical and theological issues, *CTNS Bulletin*, The Center for Theology and the Natural Sciences **21** (2), 3-9 (Spring 2001). [This paper was presented at the Sixth Conference on Chemical Evolution and first appeared in *The First Steps of Life in the Universe*, J. Chela-Flores, T. Owen and F. Raulin (eds.), Kluwer Academic Publishers, Dordrecht, pp. 365-374.

23. Davis, J. J. (1997) The search for extraterrestrial intelligence and the Christian doctrine of redemption, *Sci. Christian Belief* **9**, 21-34.

24. Crowe, M. J. (1997) A history of the extraterrestrial life debate, *Zygon* **32** (2), 147-162.

25. Spradley, J. L. (1998) Religion and the search for extraterrestrial intelligence, *Perspectives on Sci. Christian Faith* **50**, 194-203.

26. Miller, J. B. (2001) Cosmic questions and the relationship between science and religion, *Ann N Y Acad Sci.* **950**, 309-310.

27. Haught, J. F. (2001) Theology after contact religion and extraterrestrial intelligent life, *Ann N Y Acad Sci.* **950**, 296-308.

28. Francisco O. P. R. (1994) Possibilita' di una redenzione cosmica: implicazioni teologiche circa una suposta vita extraterrestre, in *Origini: l'universo, la vita, l'intelligenza*, F. Bertola, M. Calvani and U. Curi (eds.), Il Poligrafo, Padova, pp. 95-112.

29. Plato (1977) *Timaeus and Critias,* Penguin Classics, London, pp. 46-57.

30. Vakoch, D. A. (2000) Roman Catholic views of extraterrestrial intelligence: Anticipating the future by examining the past, in *When SETI Suceeds: The Impact of High-Information Contact*, Tough, A. (ed.), Bellevue, Washington, pp. 165-174.

31. Bertola, F. (2001) The plurality of worlds, in *First Steps in the Origin of Life in the Universe*, J. Chela-Flores, T. Owen and F. Raulin, (eds.), Kluwer Academic Publishers, Dordrecht, pp. 401-407.

32. Peters, T. (1994) Exo-theology: Speculations on extra-terrestrial life, *CTNS Bulletin*, The Center for Theology and the Natural Sciences, **14** (3) (Summer).

33. Milne, E. A. (1952) *Modern Cosmology and the Christian Idea of God*, Edward Cadbury Lectures, 1950, Clarendon Press, Oxford, p. 153.

34. St. Augustine (1984) *Concerning the City of God Against the Pagans*, Penguin Classics Book, London, XVII, Chapter 3.

35. Tillich, P. (1957) *Systematic Theology*, Vol. **2**, Chicago, University of Chicago Press, Chicago, p. 96.

36. Mascall, E. L. (1956) *Christian Theology and Natural Science*, Ronald Press, New York, pp. 37-39.

37. McMullin, E. (2000) Life and intelligence far from Earth: Formulating theological issues, in *Many Worlds*: *The New Universe, Extraterrestrial Life and the Theological Implications*, S. Dick (ed.), Templeton Foundation Press, Philadelphia.

38. Teilhard de Chardin, P. (1965) *The Phenomenon of Man*, Fontana Books, London.

39. Medawar, P. (1984) *The Limits of Science*, Oxford University Press, London.

40. Chela-Flores, J. (1998) The phenomenon of the eukaryotic cell, in *Evolutionary and Molecular Biology*: *Scientific Perspectives on Divine Action*, R. J., Russell, W. R. Stoeger, S. J. and F. J. Ayala, (eds.), Vatican City State/Berkeley, California, Vatican Observatory and CTNS, pp. 79-99.

41. Tracy, T. F. (1998) Evolution, divine action, and the problem of evil, in *Evolutionary and Molecular Biology*: *Scientific Perspectives on Divine Action*, R. J., Russell, W. R. Stoeger, S. J. and F. J. Ayala, (eds.), Vatican City State/Berkeley, California, Vatican Observatory and CTNS, pp. 511-530.

42. Haught, J. F. (1998) Darwin's gift to theology, in *Evolutionary and Molecular Biology: Scientific Perspectives on Divine Action*, R. J., Russell, W. R. Stoeger, S. J. and F. J. Ayala, (eds.), Vatican City State/Berkeley, California: Vatican Observatory and CTNS, pp. 393-418.

43. Ellis, G. F. R. (1998) The thinking underlying the new "scientific world-views", in *Evolutionary and Molecular Biology: Scientific Perspectives on Divine Action,* R. J., Russell, W. R. Stoeger, S. J. and F. J. Ayala, (eds.), Vatican City State/Berkeley, California, Vatican Observatory and the Center for Theology and the Natural Sciences, p. 276.

44. Plantinga, A. (1993) *Warrant*: *The Current Debate*, Oxford University Press, New York.

45. Plantinga, A (1993) *Warrant and Proper Function*, Oxford University Press, New York.

46. Plantinga, A. (2000a) *Warranted Christian Belief,* Oxford University Press, New York.

47. Plantinga, A. (2000b) Arguments for the existence of God, in *The Routledge Encyclopedia of Philosophy*, Routledge, New York.

Chapter 5

1. Carr, B. J. and Rees, M. J. (2003) Fine-tuning in living systems, *Int. J. Astrobiol.* **2** (2), 1-8.

2. Chela-Flores, J. (2001) *The New Science of Astrobiology from Genesis of the Living Cell to Evolution of Intelligent Behavior in the Universe*, Kluwer Academic Publishers, Dordrecht, The Netherlands.

3. Salam, A. (1990) *Ideals and Reality*, 3rd edition. C. H. Lai (ed.), World Scientific, Singapore, p. 282.

4. Chapman, A. (2001) *Gods in the Sky Astronomy from the Ancients to the Renaissance*, Channel 4 Books, London.

5. Bertola, F. (2001) The plurality of worlds, in *The First Steps of Life in the Universe*. J. Chela-Flores, T. Owen and F. Raulin (eds.), Kluwer Academic Publishers, Dordrecht, The Netherlands, pp. 401-407.

6. Ricci, S. (2000) *Giordano Bruno nell'Europa del Cinquecento*, Salerno Editrice, Roma.

7. Caldwell, R. and Kamionkowski, M. (2001) Echoes from the Big Bang, *Scientific American*, January, pp. 38-43.

8. Quinn, H. R. and Nir, Y. (2008) *The Mystery of the Missing Antimatter*, Princeton University Press, Princeton, N. J.

Chapter 6

1. For a more detailed account of stellar evolution with references to the literature the reader should consult: Hoyle, F. (1975) *Astronomy and Cosmology: A Modern Course*, W. H. Freeman and Company, San Francisco.

2. The Stardust cometary materials now join a collection of charged particles from the Sun gathered by NASA's Genesis mission and returned to Earth in 2004.

3. IDPs are discussed in *Lectures in Astrobiology*, M. Gargaud *et al.* (eds.), Study Edition in two volumes, Springer-Verlag, Berlin, 2006.

4. Brownlee, D. E. and Sandford, S. A. (1992) Cosmic dust, in *Exobiology in Solar System Exploration*, G. C. Carle, D. E. Schwartz and J. L. Huntington (eds.), NASA Publication SP 512, pp. 145-157.

5. Schidlowski, M. (1995) Early terrestrial life: Problems of the oldest record, in *Chemical Evolution: Self-Organization of the Macromolecules of Life*, J. Chela-Flores, M. Chadha, A. Negron-Mendoza, and T. Oshima (eds.), A. Deepak Publishing, Hampton, Virginia, USA, pp. 65-80.

6. Sleep, N. H., Zahnle, K. J., Kasting, J. F. and Morowitz, H. J. (1989) Annihilation of ecosystems by large asteroid impacts on the early Earth, *Nature* **342**, 139-142.

7. Amabile-Cuevas, C. F. and Chicurel, M. E. (1993) Horizontal gene transfer, *American Scientist* **81**, 332-341.

8. For a recent discussion of the role of HGT in evolution, we refer the reader to Pennisi, E. (1999) Is it time to uproot the tree of life? *Science* **284**, 1305-1307.

9. Woese, C. R., Kandler, O. and Wheelis, M. L. (1990) Towards a natural system of organisms. Proposal for the domains Archaea, Bacteria, and Eucarya, *Proc. Natl. Acad. Sci. USA* **87**, 4576-4579.

10. Han, T.-M. and Runnegar, B. (1992) Megascopic eukaryotic algae from the 2.1-billion-year-old Negaunee iron-formation, Michigan, *Science* **257**, 232-235.

Chapter 7

1. This chapter is based on the following papers, where several additional references to the original literature are provided:

    Messerotti, M. and Chela-Flores, J. (2007) Signatures of the ancient Sun constraining the early emergence of life on Earth, in *Space Weather. Research Towards Applications in Europe*, J. Lilensten (ed.), Springer, Dordrecht, The Netherlands, pp. 49-59 and

    Chela-Flores, J. Jerse, G., Messerotti, M. and Tuniz, C. (2008) Astronomical and astrobiological imprints on the fossil records. A review, in *From Fossils to Astrobiology*, J. Seckbach (ed.), Springer, Dordrecht, The Netherlands.

2. Muñoz Caro, G. M., Meierhenrich, U. J., Schutte, W., Barbier, B., Arcones Segovia, A., Rosenbauer, H., Thiemann, W., Brack, A., Greenberg, J. M. (2002) Amino acids from ultraviolet irradiation of interstellar ice analogues, *Nature* **416**, 403-406, and

    Bernstein, M. P., Dworkin, J. P., Sandford, S. A., Cooper, G. W. and Allamandola, L. J. (2002) Racemic amino acids from the ultraviolet photolysis of interstellar ice analogues, *Nature* **416**, 401-403.

3. Cronin, J. R. and Chang, S. (1993) Organic matter in meteorites: Molecular and isotopic analyses of the Murchison meteorite, in *The chemistry of life's origins*, J. M. Greenberg, C. X. Mendoza-Gomez and V. Pirronello (eds.), Kluwer Academic Publishers, Dordrecht, pp. 209-258.

4. Ehrenfreund, P., Glavin, D. P., Botta, O., Cooper, G. and Bada, J. L. (2001) Extraterrestrial amino acids in Orgueil and Ivuna: Tracing the parent body of CI type carbonaceous chondrite, *Proc. Natl. Acad. Sci. USA*, **98**, 2138-2141.

5. Hoppe, P., Strebel, R., Eberhadt, P., Amari, S. and Lewis, R. S. (1997) Type II supernova matter in a silicon carbide grain from the Murchison meteorite, *Science* **272**, 1314-1317.

6. Lal, D and Ligenfelter, R. E. (1991) History of the Sun during the past 4.5 Gyr as revealed by studies of energetic solar particles recorded in extraterrestrial and terrestrial samples, in *The Sun in Time,* C. P. Sonett, M. S. Giampappa and M. S. Matthews (eds.), The University of Arizona, Tucston, pp. 221-231.

7. Kerridge, J. F., Signer, P, Wieler, R., Becker, R. H., and Pepin, R. O. (1991) Long term changes in composition of solar particles implanted in extraterrestrial materials, in *The Sun in Time,* C. P. Sonett, M. S. Giampappa and M. S. Matthews (eds.), The University of Arizona, Tucston, pp. 389-412.

8. Matsui, T. and Abe, Y. (1986) Evolution of an impact-induced atmosphere and magma ocean on the accreting Earth, *Nature* **319**, 303-305.

9. Hunten, D. M. (1993) Atmospheric evolution of the terrestrial planets, *Science* **259**, 915-920.

10. Kasting, J. F. (1993) Earth's early atmosphere, *Science* **259**, 920-926.

11. Kerridge, J. F. (1975) Solar nitrogen: Evidence for a secular increase in the ratio of nitrogen-15 to nitrogen-14, *Science* **188**, 162-164.

12. Wieler, R., Humbert, F. and Marty, B. (1999) Evidence for a predominantly non-solar origin of nitrogen in the lunar regolith revealed by single grain analyses, *Earth Planet. Sci. Lett*, **167**, 47-60.

13. Ozima, M., Seki, K., Terada, N., Miura, Y. N., Podosek, F. A. and Shinagawa, H. (2005) errestrial nitrogen and noble gases in lunar soils, *Nature* **436**, 655-659.

14. Oparin, A. I. (1953) *Origin of Life*, Dover, New York.

15. Ponnamperuma, C. and Chela-Flores, J. (eds.) (1995) *Chemical Evolution: The Structure and Model of the First Cell. The Alexander Ivanovich Oparin 100th Anniversary Conference*, Kluwer Academic Publishers, Dordrecht, The Netherlands.

16. Messerotti, M. (2004) Space weather and space climate, in *Life in the Universe From the Miller Experiment to the Search for Life on Other Worlds.* J. Seckbach, J. Chela-Flores, T. Owen, F. Raulin (eds.), Series: Cellular Origin, Life in Extreme Habitats and Astrobiology, pp. 177-180.

17. Goswami, J. N. (1991) Solar flare heavy-ion tracks in extraterrestrial objects, in *The Sun in Time,* C. P. Sonett, M. S. Giampappa and M. S. Matthews (eds.), The University of Arizona, Tucston, pp. 426-444.

18. Canuto, V. M., Levine, J. S., Augustsson, T. R. and Imhoff, C. L. (1982) Oceanic protection of prebiotic organic compounds from UV radiation, *Nature* **296**, 816-820.

19. Canuto, V. M., Levine, J. S., Augustsson, T. R., Imhoff, C. L. and Giampapa, M. S. (1983) The young Sun and the atmosphere and photochemistry of the early Earth, *Nature* **305**, 281-286.

20. Thomas, C. H., Jackman, A. L., Melott, C. M., Laird, R. S., Stolarski, N., Gehrels, J. K. Cannizzo, and Hogan, D. P. (2005) *Astrophy. J. Lett.* **622**, L153.

21. Renard, P., Koeck, C., Kemble, S., Atzei, A. and Falkner, P. (2005). System concepts and enabling technologies for an ESA low-cost mission to Jupiter/Europa, (*Proceedings of 55th International Astronautical Congress*), Vancouver, Canada, 2004.

22. Greenberg, R. (2005) *Europa the Ocean Moon Search for an Alien Biosphere*, Springer, Berlin.

23. Humes, D. H., Alvarez, J. M., O'Neal, R. L. and Kinard, W. H. (1974) The interplanetary and near-Jupiter meteoroid environments, *J. Geophys. Res.* **79** (25), 3677-3684.

24. Srama, T., Ahrens, J., Altobelli, N., Auer, S., Bradley, J. G., Burton, M., Dikarev, V. V., Economou, T., Fechtig, H., Görlich, M., Grande, M., Graps, A., Grün, E., Havnes, O., Helfert, S., Horanyi, M., Igenbergs, E., Jessberger, E. K., Johnson, T. V., Kempf, S., Krivov, A. V., Krüger, H., Mocker-Ahlreep, A., Moragas-Klostermeyer, G., Lamy, P., Landgraf, M., Linkert, D., Linkert, G., Lura, F., McDonnell, J. A. M., Möhlmann, D., Morfill, G. E., Müller, M., Roy, M., Schäfer, G., Schlotzhauer, G., Schwehm, G. H., Spahn, F., Stübig, M., Svestka, J., Tschernjawski, V., Tuzzolino, A. J., Wäsch and R., Zook, H. A. (2004) The Cassini cosmic dust analyzer, *Space Sci. Rev.* **114**, 465-518.

25. Grun, E., Zook, H. A., Baguhl, M., Balogh, A., Bame, S. J., Fechtig, H., Forsyth, R., Hanner, M. S., Horanyi, M., Kissel, J., Lindblad, B.-A., Linkert, D., Linkert, G., Mann, I., McDonnell, J. A. M., Morfill, G. E., Phillips, J. L., Polanskey, C., Schwehm, G., Siddique, N., Staubach, P., Svestka, J. and Taylor, A. (1993) Discovery of Jovian dust streams and interstellar grains by the Ulysses spacecraft, *Nature* **362**, 428-430.

26. Grun, E., Hamilton, D. P., Riemann, R., Dermott, S., Fechtig, H., Gustafson, B. A., Hanner, M. S., Heck, A., Horanyi, M., Kissel, J., Kruger, H., Lindblad, B.-A., Linkert, D., Linkert, G., Mann, I., McDonnell, J. A. M., Morfill, G. E., Polanskey, C., Schwehm, G., Srama, R. and Zook, H. A. (1996) Dust measurements during the initial Galileo Jupiter approach and Io encounter, *Science* **274**, 399-401.

27. Showalter, M. R., Burns, J. A., Cuzzi, J. N., and Pollack, J. B. (1985) Discovery of Jupiter's "gossamer" ring, *Nature* **316**, 526-528.

28. Zook, H. A., Grun, E., Baguhl, M., Hamilton, D. P., Linkert, G., Liou, J.-C., Forsyth, R. and Phillips, J. L. (1996) Solar wind magnetic field bending of Jovian dust trajectories, *Science* **274**, 1501-1503.

29. Grün, E., Krüger, H., Graps, A., Hamilton, D. P., Heck, A., Linkert, G., Zook, H. A., Dermott, S., Fechtig, H.,Gustafson, B. A., Hanner, M. S., Horányi, M., Kissel, J., Lindblad, B. A., Linkert, D., Mann, I., McDonnell, J. A. M., Morfill, G. E., Polanskey, C., Schwehm, G., and Srama, R. (1998) Galileo observes electromagnetically coupled dust in the jovian magnetosphere, *J. Geophys. Res.* **103**, 20011-20022.

30. Graps, A. L., Grun, E., Svedhem, H., Kruger, H., Horanyi, M., Heck, A. and Lammers, S. (2000) Io as a source of the jovian dust streams, *Nature* **405**, 48-50.

31. Chela-Flores, J. (2006) The sulphur dilemma: Are there biosignatures on Europa's icy and patchy surface? *Int. J. Astrobiol.* **5**, 17-22, Cambridge University Press, http://www.ictp.it/~chelaf/ss64.html.

Chapter 8

1. Conway Morris, S. (2003) *Life's Solution Inevitable Humans in a Lonely Universe,* Cambridge University Press.

2. Chela-Flores, J. (2001) *The New Science of Astrobiology from Genesis of the Living Cell to Evolution of Intelligent Behavior in the Universe,* Kluwer Academic Publishers, Dordrecht, The Netherlands.

3. Akindahunsi, A. A. and Chela-Flores, J. (2004) On the question of convergent evolution in biochemistry, in *Life in the Universe: From the Miller Experiment to the Search for Life on Other Worlds,* J. Seckbach, J. Chela-Flores, T. Owen and F. Raulin (eds.), Kluwer Academic Publishers, Dordrecht, pp. 135-138.

4. Chela-Flores, J. (2007) Fitness of the cosmos for the origin and evolution of life: From biochemical fine-tuning to the Anthropic Principle, in *Fitness of the Cosmos for Life: Biochemistry and Fine-Tuning,* J. D. Barrow, S. Conway Morris, S. J. Freeland and C. L. Harper, (eds.), Cambridge University Press, pp.151-166.

5. Dawkins, R. (1983) Universal darwinism, in *Evolution from Molecules to Men,* D. S. Bendall (ed.), Cambridge University Press, London, pp. 403-425.

6. Pace, N. R. (2001) The universal nature of biochemistry, *Proc. Natl. Acad. Sci. USA* **98**, 805-808.

7. Chela-Flores, J. (2007) Testing the universality of biology, *Int. J. Astrobiol.* **6** (3), 241-248. Cambridge University Press.

8. Tudge, C. (1991) *Global Ecology,* Natural History Museum Publications, London, p. 67.

9. Nigel-Hepper, F. (1982) *Kew: Gardens for Science and Pleasure,* Her Majesty's Stationary Office, London, p. 81.

10. Fontana, W. and Buss, L. W. (1994) What would be conserved if "the tape were played twice"? *Proc. Natl. Acad. Sci. USA* **91**, 757-761.

11. Conway Morris, S. (1998) *The Crucible of Creation. The Burgess Shale and the Rise of Animals,* Oxford University Press, London, p. 202.

12. Chen, L., DeVries, A. L. and Cheng, C-H. C. (1997) Convergent evolution of antifreeze glycoproteins in Antarctic notothenioid fish and Arctic cod. *Proc. Natl. Acad. Sci. USA* **94**, 3817-3822.

13. Yokoyama, R. and Yokoyama, S. (1990) Convergent evolution of the red- and green-like visual pigment genes in fish, *Astynax fasciatus,* and human, *Proc. Natl. Acad. Sci. USA* **87**, 9315-9318.

14. Austin, D. F. (1998) Parallel and convergent evolution in the Convolvulaceae, in *Diversity and Taxonomy of Tropical Flowering Plants,* P. Mathews and M. Sivadasan (eds.), Mentor Books, Calicut, India, pp. 201-234.

15. Doolittle, R. F. (1994) Convergent evolution: The need to be explicit, *Trends Biochem. Sci.* **19**, 15-18.

16. Harper, E. M. Taylor, J. D. and Crame, J. A. (2000) *The Evolutionary Biology of the Bivalvia,* Geological Society Special Publication, Vol. **177**, London.

17. Shadwick, R. E. (2005) How tunas and lamnid sharks swim: An evolutionary convergence, *Amer. Scientist* **93** (6), 524.

18. Bernal, D., Donley, J. M., Shadwick, R. E. and Syme, D. A. (2005) Mammal-like muscles power swimming in a cold-water shark, *Nature* **437**, 1349-1352.

Chapter 9

1. Krubitzer, L. (1995) The organization of neocortex in mammals: are species differences really so different? *Trends in Neuroscience* **18**, 408-417.

2. Dawkins, R. (1983) Universal darwinism, in *Evolution from Molecules to Men*, D. S. Bendall (ed.), Cambridge University Press, London, pp. 403-425.

3. Chela-Flores, J. (2001) *The New Science of Astrobiology from Genesis of the Living Cell to Evolution of Intelligent Behavior in the Universe*, Kluwer Academic Publishers, Dordrecht, The Netherlands.

4. Chela-Flores, J. and Kumar, N. (2008) Returning to Europa: Can traces of surficial life be detected? *Int. J. Astrobiol.*, in press.

5. The reader is referred to the following paper for a more extensive bibliography: Villegas, R, Castillo, C. and Villegas, G. M. (2000) The origin of the neuron: The first neuron in the phylogenetic tree of life, in *Astrobiology from the Big Bang to Civilisation,* J. Chela-Flores, G. A. Lemarchand and J. Oro (eds.), Kluwer Academic Publishers, Dordrecht, The Netherlands. pp. 195-211.

6. Chela-Flores, J. (1998) A search for extraterrestrial eukaryotes: Physical and biochemical aspects of exobiology, *Origins Life Evol. Biosphere* **28**, 583-596.

7. Chela-Flores, J. (2000) Testing the Drake equation in the solar system, in *A New Era in Astronomy*, G. A. Lemarchand and K. Meech (eds.), ASP Conference Series, San Francisco, Vol. **213**, pp. 402-410.

8. Fortes, A. D. (2000) Exobiological implications of a possible ammonia-water ocean inside Titan, *Icarus* **146**, 444-452.

9. Chela-Flores, J. (1996) Habitability of Europa: Possible degree of evolution of Europan biota, in *Europa Ocean Conference at San Juan Capistrano Research Institute*, San Juan Capistrano, California, USA, 12-14 November, pp. 21-21a.

10. Horvath, J., Carsey, F., Cutts, J., Jones, J. Johnson, E., Landry, B., Lane, L., Lynch, G., Chela-Flores, J., Jeng, T-W. and Bradley, A. (1997) Searching for ice and ocean biogenic activity on Europa and Earth, *Instruments, Methods and Missions for Investigation of Extraterrestrial Microorganisms,* R. B. Hoover (ed.), Proc. SPIE, Vol. **3111**, pp. 490-500.

11. Doran, P. T., Stone, W., Priscu, J., McKay, C., Johnson, A. and Chen, B. (2007) *Environmentally Non-Disturbing Under-Ice Robotic ANtarctiC Explorer (ENDURANCE)*, American Geophysical Union, Fall Meeting 2007, abstract #P52A-05.

12. Chyba, C. (2000) Energy for microbial life on Europa, *Nature* **403**, 381-382.

13. Schulze-Makuch, D. and Irwin, L. N. (2002) Energy cycling and hypothetical organisms in Europa's Ocean, *Astrobiology* **2**, 105-121.

14. Zolotov, M. Y. and Shock, E. L. (2003) On energy for biologic sulfate reduction in a hydrothermally formed ocean on Europa, *J. Geophys. Res.*, **108** (E4), 5022, doi:10.1029/2002JE001966.

15. Chela-Flores, J. (2006) The sulphur dilemma: Are there biosignatures on Europa's icy and patchy surface? *Int. J. Astrobiol.* **5**, 17-22.

16. Bruno, G. (2000) *De l'infinito, universo e mondi*, Venice, 1584. [English translation: On the infinite universe and innumerable worlds, Cambridge, 1650]. For a more precise bibliographic reference*: Giordano Bruno 1548-1600*, Biblioteca di Bibliografia Italiana Vol. **164**, Roma, Leo S. Olschki Editore, pp. 105-106.

17. Russell, R. (2001) Life in the universe: Philosophical and theological issues*, CTNS Bulletin*, The Center for Theology and the Natural Sciences **21**(2), Spring 2001.

Chapter 10

1. Monod, J. (1971) *Chance and Necessity: An Essay on the Natural Philosophy of Modern Biology.* Alfred A. Knopf, New York.

2. De Duve, C. (1995) *Vital Dust. Life as a Cosmic Imperative.* Basic Books, New York.

3. De Duve, C. (2002) *Life Evolving Molecules Mind and Meaning*, Oxford University Press, New York.

4. De Duve, C. (2005) *Singularities Landmarks on the Pathway of Life*, Cambridge University Press, London.

5. Knoll, A. H. (1995) Life story, *Nature* **375**, 201-202.

6. Szathmary, E. (2002) The gospel of inevitability was the universe destined to lead to the evolution of humans? *Nature* **419**, 779-780.

7. Foote, M. (1998) Contingency and convergence, *Science* **280**, 2068-2069.

8. Erwin, D. H. (2003) The Goldilocks hypothesis, *Science* **302**, 1682-1683.

9. Penny, D. (2006) Defining moments, *Nature,* **442**, 745-746.

10. Gazzaniga, M. S., Ivry, R. B. and Mangun, G. R. (1998) *Cognitive Neuroscience. The Biology of the Mind*, W. W. Norton & Company, New York, pp. 590-593. (The chapter concerned, "Evolutionary Perspectives" was written in collaboration with Leah Krubitzer.)

11. Krubitzer, L. (1995) The organization of neocortex in mammals: Are species differences really so different? *Trends Neurosci.* **18**, 408-417.

12. Manger, P., Sum, M., Szymanski, M., Ridgway, S. and Krubitzer, L. (1998) Modular ubdivisions of dolphin insular cortex: Does evolutionary history repeat itself? *J.Cognitive Neurosci.* **10**, 153-166.

13. Krubitzer, L. and Kahn, D. M. (2003) Nature versus nurture revisited: An old idea with a new twist, *Prog Neurobiol* **70**, 33–52.

14. Weinreich, D. M., Delaney, N. F., DePristo, M. A. and Hartl, D. L. (2006) Darwinian evolution can follow only very few mutational paths to fitter proteins, *Science* **312**, 111-113.

15. Ziegler B. (1983) *Introduction to Palaeobiology: General Palaeontology*, Ellis Horwood Limited, Chichester, UK, 225 pp.

16. Chela-Flores, J. (2001) *The New Science of Astrobiology from Genesis of the Living Cell to Evolution of Intelligent Behavior in the Universe*, Kluwer Academic Publishers, Dordrecht, The Netherlands, pp. 149-156.

17. Chela-Flores, J. (2007) Fitness of the cosmos for the origin and evolution of life: From biochemical fine-tuning to the Anthropic Principle, in *Fitness of the Cosmos for Life*: *Biochemistry and Fine-Tuning*, J. D. Barrow, S. Conway Morris, S. J. Freeland and C. L. Harper (eds.), Cambridge University Press, pp. 151-166.

18. Austin, D. F. (1998) Parallel and convergent evolution in the Convolvulaceae, in *Diversity and Taxonomy of Tropical Flowering Plants,* P. Mathews and M. Sivadasan (eds.), Mentor Books, Calicut, India, pp. 201-234.

19. Tramontano, A. (2002) Private communication.

20. Doolittle, R. F. (1994) Convergent evolution: The need to be explicit, *Trends Biochem. Sci.* **19**, 15-18.

21. Kornegay, J. (1996) Molecular genetics and evolution of stomach and nonstomachlysozymes in the hoatzin, *J. Mol. Evol.* **42**, 676-684.

22. Zhang, J. and S. Kumar (1997) Detection of convergent and parallel evolution at the amino acid sequence level, *Mol. Biol. Evol.* **14**, 527-536.

23. Cronin, J. R. and Chang, S. (1993) Organic matter in meteorites: Molecular and isotopic analyses of the Murchison meteorite, in *The Chemistry of Life's Origins*, J. M. Greenberg, C. X. Mendoza-Gomez and V. Pironello (eds.), Kluwer Academic Publishers, Dordrecht, pp. 209-58.

24. Greenberg, J. M. and Mendoza-Gomez, C. X. (1993) Interstellar dust evolution: A Reservoir of prebiotic molecules, in *The Chemistry of Life's Origins*, Greenberg, J. M., Mendoza-Gomez, C. X. and Pironello. V. (eds.), Kluwer Academic Publishers, Dordrecht, The Netherlands, pp. 1-32.

25. Hoppe, P., Strebel, R., Eberhadt, P., Amari, S. and Lewis, RS (1997) Type II supernova matter in a silicon carbide grain from the Murchison meteorite, *Science* **272**, 1314-17.

26. Kerr, R. A. (2002) Winking star unveils planetary birthplace, *Science* **296**, 2312-13.

27. Schneider, G., Smith, B. A., Becklin, E. E., Koerner, D. W., Meier, R., Hines, D. C., Lowrance,, P. J., Terrile, R. J., Thompson, R. I., and Rieke, M. (1999) NICMOS imaging of the HR 4796A circumstellar disk, *Astrophys. J.* **513**, L1217-30.

28. Delsemme A. H. (2000) Cometary origin of the biosphere, *Icarus* **146**, 313-325.

29. Ekers, R. D., Cullers, D. K., Billingham, J. and Scheffer, L. K. (eds.) (2002) *SETI 2020*, SETI Press, Mountain View California, USA.

30. Ponnamperuma, C. and Chela-Flores, J. (eds.) (1995) *Chemical Evolution: The Structure and Model of the First Cell*, Kluwer Academic Publishers, Dordrecht, The Netherlands.

31. Ponnamperuma, C. and Chela-Flores, J. (eds.). (1993) *Chemical Evolution: Origin of Life,* A. Deepak Publishing, Vol. **135**, Hampton, Virginia, USA.

32. Chela-Flores, J., Chadha, M., Negron-Mendoza, A. and Oshima, T. (eds.) (1995) *Chemical Evolution: Self-Organization of the Macromolecules of Life.* A. Deepak Publishing, Vol. **139**, Hampton, Virginia, USA.

33. Chela-Flores, J. and Raulin, F. (eds.). (1996*) Chemical Evolution: Physics of the Origin and Evolution of Life* (The Cyril Ponnamperuma Memorial Conference), Kluwer Academic Publishers, Dordrecht, The Netherlands.

34. Chela-Flores, J. and Raulin, F. (eds.) (1998) *Exobiology: Matter, Energy, and Information in the Origin and Evolution of Life in the Universe*, Kluwer Academic Publishers, Dordrecht, The Netherlands.

35. Chela-Flores, J., Lemarchand, G. A. and Oro, J. (2000) *Astrobiology: Origins from the Big Bang to Civilisation.* Kluwer Academic Publishers, Dordrecht, The Netherlands.

36. Chela-Flores, J, Owen, T. and Raulin, F. (2001) *The First Steps of Life in the Universe*, Kluwer Academic Publishers, Dordrecht, The Netherlands.

37. Seckbach, J. Chela-Flores, J. Owen, T. Raulin, F. (eds.) (2004*) Life in the Universe from the Miller Experiment to the Search for Life on Other Worlds*, Kluwer Academic Publishers, Dordrecht, The Netherlands.

38. Chela-Flores, J. (2006) The sulphur dilemma: Are there biosignatures on Europa's icy and patchy surface? *Int. J. Astrobiol.* **5**, 17-22.

39. McCord, T. B., Hansen, G. B., Clark, R. N., Martin, P. D., Hibbitts, C. A., Fanale, F. P., Granahan, J. C., Segura, N. M., Matson, D. L., Johnson, T. V., Carlson, R. W., Smythe, W. D., Danielson, G. E. and the NIMS team (1998) Non-water-ice constituents in the surface material of the icy Galilean satellites from the Galileo near-infrared mapping spectrometer investigation, *Geophys. Res.* **103**(E4), pp. 8603-8626.

40. Singer, E. (2003) Vital clues from Europa, *New Scientist Magazine*, issue No. 2414 (27 September), p. 23, http://www.newscientist.com/contents/issue/2414.html (Option: "Vital clues from Europa").

41. Bhattacherjee, A. B and Chela-Flores, J. (2004) Search for bacterial waste as a possible signature of life on Europa, in *Life in the Universe*, J. Seckbach, J. Chela-Flores, T. Owen and F. Raulin (eds.), Cellular Origin and Life in Extreme Habitats and Astrobiology, Vol. **7**, Springer, Dordrecht, The Netherlands, pp. 257-260.

42. Aretxaga, R. (2004) Astrobiology and biocentrism, in *in Life in the Universe*, J. Seckbach, J. Chela-Flores, T. Owen and F. Raulin (eds.), Cellular Origin and Life in Extreme Habitats and Astrobiology, Vol. **7**, Springer, Dordrecht, The Netherlands, pp. 345-348.

43. Thomson, R. E. and Delaney, J. R. (2001) Evidence for a weakly stratified Europan ocean sustained by seafloor heat flux, *Jour. Geophys. Res.* **106**(E6) 12 355-12 365.

Chapter 11

1. Little, C. T. S., Herrington, R. J., Maslennikov, V. V., Morris, N. J. and Zaykov, V. V. (1997) Silurian hydrothermal-vent community from the southern Urals, Russia, *Nature* **385**, 146-148.

2. Little, C. T. S., Herrington, R. J., Maslennikov, V. V. and Zaykov, V. V. (1998) The fossil record of hydrothermal vent communities, in *Modern Ocean Floor Processes and the Geologic Record*, R. A. Mills and K. Harrison (eds.), Geological Society, London, pp. 259-270.

3. Jacobs, D. K. and Lindberg, D. R. (1998) Oxygen and evolutionary patterns in the sea: Onshore/offshore trends and recent recruitment of deep-sea faunas, *Proc. Natl. Acad. Sci. USA* **95**, 9396- 9401.

4. Ekers, R. D., Kent Cullers, D., Billingham, J. and Scheffer, L. K. (eds.), (2002) *SETI 2020*, SETI Press, Mountain View CA.

5. Chela-Flores, J. (2003) Testing evolutionary convergence on Europa, *Int. J. Astrobiol.* **2** (4), 307-312 (Cambridge University Press).

6. Huey, R., Gilchrist, G., Carlson, M., Berrigan, D. and Serra, L. (2000) Rapid evolution of a geographic cline in size in an introduced fly, *Science* **287**, 308-309.

7. Losos, J. B., Jackman, T.R., Larson, A., de Queiroz, K. and Rodriguez-Schettino, L. (1998) Contingency and determinism in replicated adaptive radiations of island lizards, *Science* **279**, 2115-2118.

8. Vogel, G. (1998) For island lizards, history repeats itself, *Science* **279**, 2043.

9. De Duve, C. (1995) *Vital Dust. Life as a Cosmic Imperative*, Basic Books, A Division of Harper-Collins Publishers, New York, pp. 296-297.

10. De Duve, C. (2002) *Life Evolving Molecules Mind and Meaning*, Oxford University Press, New York.

Chapter 12

1. A preliminary version of the ideas expressed in this chapter was written in the paper: Astrobiological reflections on faith and reason, the issues of agnosticism, relativism and natural selection, J. Seckbach (ed.) (2008), World Scientific Publishers, to appear, and in reference 13 below.

2. Barbour, I. G. (1995) Historical and contemporary relations in science and religion, in *Physics, Philosophy and Theology. A Common Quest for Understanding*, 2nd edition, R. J. Russell, W. R. Stoeger S. J. and G. V. Coyne S. J. (eds.), Vatican Observatory Foundation, Vatican City State, pp. 21-48.

3. Einstein, A. (1950) *Out of My Later Years*, Philosophical Library, New York, pp. 21-37.

4. Augustine of Hippo (1984) *Concerning the City of God Against the Pagans*, Penguin Classics, London (cf. Book XVII, p. 4).

5. O'Meara, J. J. (1984) Introduction to St. Augustine's *Concerning the City of God Against the Pagans,* Penguin Classics, London, pp. 1-15.

6. Desmond, A. and Moore, J. (1991) *Darwin*, (Chapter 38—Disintegrating speculations), M. Joseph, London, pp. 568-586.

7. O'Grady, P. (2002) *Relativism.* Acumen, Buks, UK.

8. Ambrosi, E. (2005) *Il bello del relativismo*, Marsilio Editore, Venice, Italy.

9. Soffen, G. A. (1976) Scientific results from the Viking Mission, *Science* **194**, 1274-1276.

10. Horneck, G. (1995) Exobiology, the study of the origin, evolution and distribution of life within the context of cosmic evolution: a review, *Planet. Space Sci.* **43**, 189-217.

11. Richardson, J. D., Kasper, J. C., Wang, C., Belcher, J. W. and Lazarus, A. J. (2008) Cool heliosheath plasma and deceleration of the upstream solar wind at the termination shock, *Nature* **454**, 63-66.

12. Blanc, M. and the LAPLACE consortium (2008) LAPLACE: A mission to Europa and the Jupiter System, *Astrophysical Instruments Methods,* in press.

13. Chela-Flores, J. (2005) Fitness of the universe for a second Genesis: Is it compatible with science and Christianity? *Sci. Christian Belief* **17** (2), 187-197.

14. Schopf, J. W. (1993) Microfossils of the Earth Archaean Apex Chert: New evidence of the antiquity of life, *Science* **260**, 640-646.

15. Brasier, M. D., Green, O. W., Jephcoat, A. P., Kleppe, A. K., Van Kranendonk, M. J., Lindsay, J. F., Steele, A. and Grassineau, N. V. (2002) Questioning the evidence for Earth's oldest fossils, *Nature* **416**, 76-81.

16. Schonborn, C. (2005) Finding design in nature, New York Times, July 7.

17. Coyne S. J., G. (2005) God's chance creation, *The Tablet* **6**, August.

18. Russell, B. (1991) *History of Western Philosophy and Its Connection with Political and Social Circumstances from the Earliest Times to the Present Day*, Routledge, London, p. 13.

19. Haught, J. F. (1998) Darwin's gift to theology, in R. J. Russell, W. R. Stoeger S. J. and F. J. Ayala, (eds.), *Evolutionary and Molecular Biology: Scientific Perspectives on Divine Action*, Vatican City State/Berkeley, California, Vatican Observatory and CTNS, pp. 393-418.

20. Haught, J. F. (2005) The Boyle Lecture 2003: Darwin, design and the promise of nature, *Sci. Christian Belief* **17** (1), 5-20.

21. Russell, R. J., N. Murphy and C. J. Isham, (eds.) (1996) *Quantum Cosmology and the Laws of Nature Scientific Perspectives on Divine Action*, 2nd edition, Vatican Observatory, Vatican City State, (cf., Russell's Introduction, pp. 1-31).

Chapter 13

1. This chapter is based on preliminary contributions in Seckbach, J. and Gordon, R. (eds.) *Divine Action and Natural Selection*, Singapore, WSP.

2. NAS (1999) *Science and Creationism: A View from the National Academy of Sciences*, 2nd. edition, National Academy of Sciences, p. 35.

3. Russell, B. (1991) *History of Western Philosophy and Its Connection with Political and Social Circumstances from the Earliest Times to the Present Day*, Routledge, London, p. 13.

4. Haught, J. F. (2005) The Boyle Lecture 2003: Darwin, design and the promise of nature, *Sci. Christian Belief* **17**(1), pp. 5-20.

5. McMullin, E. (2000) *Life and intelligence far from Earth: Formulating theological issues,* in *Many Worlds*, S. Dick (ed.), Templeton Foundation Press, Philadelphia.

6. Peacocke, A. (1988) Biological evolution—A positive theological appraisal, in *Evolutionary and Molecular Biology: Scientific Perspectives on Divine Action,* R. J. Russell, W. R. Stoeger S. J. and F. J. Ayala (eds.), Vatican City State/Berkeley, California, Vatican Observatory and the Center for Theology and the Natural Sciences, pp. 357-376.

7. Polkinghorne. J. (1996) *Scientists as Theologians*, SPCK, London.

8. Haught, J. F. (1998) Darwin's gift to theology, in *Evolutionary and Molecular Biology: Scientific Perspectives on Divine Action*, Loc. cit. (ref 6 above), pp. 393-418.

9. Russell, B. (1995) *Religion and Science*, Dover, New York.

10. Seckbach, J. (ed.) (2004) *Origins*: *Genesis, Evolution and the Biodiversity of Life*, Cellular Origin, Life in Extreme Habitats and Astrobiology (COLE), Vol. 6, Springer, Dordrecht, The Netherlands.

11. Seckbach, J. (ed.) (2006) *Life as We Know It*, Cellular Origins, Life in Extreme Habitats and Astrobiology, Springer, Dordrecht, The Netherlands.

12. Seckbach, J., Chela-Flores, J., Owen, T. and Raulin F. (eds.) (2004) *Life in the Universe*, Kluwer Academic, Dordrecht, The Netherlands.

13. Coyne S. J., G. (1998) *The concept of matter and materialism in the origin and evolution of life*, in J. Chela-Flores and F. Raulin (eds.), Loc. cit. (in ref. 21), pp. 71-80.

14. De Duve, C. (2002) *Life Evolving Molecules Mind and Meaning*. New York, OUP.

15. Chela-Flores, J. (2004) *The New Science of Astrobiology*, COLE Series, Vol. 3, Kluwer Academic Publishers, Dordrecht, The Netherlands, p. 251.

16. Einstein, A. (1950) *Out of My Later Years*, Philosophical Library, New York, p. 25.

17. Spinoza (2002) *Ethics*, Everyman, London.

18. Ponnamperuma, C. and Chela-Flores, J. (eds.) (1993) *Chemical Evolution*: *Origin of Life*, A. Deepak Publishing, Vol. 135, Hampton, Virginia, USA.

19. Ponnamperuma, C. and Chela-Flores, J. (eds.) (1995) *Chemical Evolution*: *The Structure and Model of the First Cell*, Kluwer Academic Publishers, Dordrecht, The Netherlands.

20. Chela-Flores, J. and Raulin, F. (eds.). (1996) *Chemical Evolution: Physics of the Origin and Evolution of Life*, Kluwer Academic Publishers, Dordrecht, The Netherlands.

21. Chela-Flores, J. and Raulin, F. (eds.) (1998) *Exobiology: Matter, Energy, and Information in the Origin and Evolution of Life in the Universe*, Kluwer Academic, Dordrecht, The Netherlands.

22. Chela-Flores, J, Owen, T. and Raulin, F. (2001) *The First Steps of Life in the Universe*. Kluwer Academic Publishers, Dordrecht, The Netherlands.

23. Darwin, C. (1859) *The Origin of Species by Means of Natural Selection or the Preservation of Favored Races in the Struggle for Life*, John Murray/Penguin Books, London, 1968, p.455.

24. Reznick, D. N., Shaw, F. H., Rodd, F. H., and Shaw, R. G. (1997) Evaluation of the rate of evolution in natural populations of guppies (*Poecilia reticulata*), *Science* **275**, pp.1934-1937.

25. Morell, V. (1997) Evolution: predator-free guppies take an evolutionary leap forward, *Science* **275**, 1880.

26. Eldredge, N. (2006) *Darwin*: *Discovering the Tree of Life*, Norton, New York.

27. Gould, S. J. (1999) *Rocks of Ages*, Ballantine, New York.

28. Coyne S. J., G. (2005) God's chance creation, *The Tablet* **6**, http://www.ictp.it/~chelaf/Coyne.pdf.

29. Ayala, F. J. (1998) Darwin's devolution: Design without designer, in *Evolutionary and Molecular Biology: Scientific Perspectives on Divine Action*, Loc. cit. (ref. 6 above). pp. 101-116.

30. Leibniz, G. (1714) *The Principles of Nature and Grace, Based on Reason*, in *Philosophical Papers and Letters*, L. Loemker (ed.), D. Reidel, Dordrecht, 1969, pp. 636-42.

Chapter 14

1. This chapter is based on an earlier manuscript Chela-Flores, J. (2006) Destinies of life and the universe: The final frontiers of astrobiology and cosmology, in *Life as We Know It*, Cellular Origins, Life in Extreme Habitats and Astrobiology, Springer, Dordrecht, The Netherlands, pp. 505-517.

2. McKay, C. P. (2001) The search for a second Genesis in our Solar System, in Chela-Flores, J., Owen, T. and Raulin, F. (eds.), *The First Steps of Life in the Universe*, Dordrecht, Kluwer Academic Publishers, pp. 269-277.

3. De Duve, C. (2002) *Life Evolving Molecules Mind and Meaning*. New York, Oxford University Press.

4. Dear, P. (2006) *The Intelligibility of Nature How Science Makes Sense of the World*, The University of Chicago Press, Chicago, USA, p. 254.

5. Russell, R. (2001) Life in the universe: Philosophical and theological issues, *CTNS Bulletin*, The Center for Theology and the Natural Sciences **21**, 3-9 (Spring 2001); J. Chela-Flores, T. Owen and F. Raulin (eds.) (2001) *The First Steps of Life in the Universe*, Kluwer Academic Publishers, Dordrecht, pp. 365-374.

6. Weinberg, S. (1977) *The First Three Minutes*, Fontana/Collins, London.

7. Spinoza, B. (2002) *Ethics*, Everyman, London.

8. Hegel, G. W. F. (1967) *Philosophy of Right*, Oxford University Press, London.

9. Kierkegaard, S. (1986) *Fear and Trembling*, Penguin Books, London.

10. Kierkegaard, S. (1981) *The Concept of Anxiety*, Kierkegaard's Writings, Vol. **8**, Princeton University Press, Princeton, USA.

11. Heidegger, M. (1996) *Being and Time*, A Translation of *Sein und Zeit*, 1927 (SUNY series in Contemporary Continental Philosophy), State University of New York Press, New York.

12. Sartre, J. P. (1947) *Situations*, I, Gallimard, Paris.

13. Husserl, E. (1999) *The Idea of Phenomenology*, Kluwer Academic Publishers, Dordrecht.

14. Camus, A. (1991) *The Myth of Sisyphus and Other Essays,* Vintage International, New York.

15. Weinberg, S. (1993) *Dreams of a Final Theory*, Vintage International, London.

16. Dawkins, R. (1983) Universal Darwinism, in *Evolution from Molecules to Men*, D. S. Bendall (ed.), Cambridge University Press, London, pp. 403-425.

17. Drake, F. and Sobel, D. (1992) *Is There Anyone Out There? The Scientific Search for Extraterrestrial Intelligence*, Delacorte Press, New York.

18. Ekers, R. D., Kent Cullers, D., Billingham, J. and Scheffer, L. K. (eds.), (2002) *SETI 2020*, SETI Press, Mountain View California, USA.

19. Bruno, G. (2000) *De l'infinito, universo e mondi*, Venice, 1584. [English translation: On the Infinite Universe and Innumerable Worlds, Cambridge, 1650]. For a more precise bibliographic reference: *Giordano Bruno 1548-1600*, Biblioteca di Bibliografia Italiana Vol. **164**, Leo S. Olschki Editore, Rome, Italy, pp. 105-106.

20. Akindahunsi, A. A. and Chela-Flores, J. (2004) On the question of convergent evolution in biochemistry, in *Life in the Universe from the Miller Experiment to the Search for Life on Other Worlds*, J. Seckbach, J. Chela-Flores, T. Owen and F. Raulin (eds.), Dordrecht, pp. 135-138.

21. Conway Morris, S. (2003) *Life's Solution: Inevitable Humans in a Lonely Universe*, Cambridge University Press, Cambridge UK.

22. Chela-Flores, J. (2003) Testing evolutionary convergence on Europa, *Int. J. Astrobiol.* **2**, 307-312.

23. Chela-Flores, J. (2006) The sulphur dilemma: Are there biosignatures on Europa's icy and patchy surface? *Int. J. Astrobiol.* **5**, 17-22.

24. McCord, T. B., Hansen, G. B., Clark, R. N., Martin, P. D., Hibbitts, C. A., Fanale, F. P., Granahan, J. C., Segura, N. M., Matson, D. L., Johnson, T. V., Carlson, R. W., Smythe, W. D., Danielson, G. E. and the NIMS Team (1998) Non-water-ice constituents in the surface material of the icy Galilean satellites from the Galileo near-infrared mapping spectrometer investigation, *J. Geophys. Res.* **103**(E4), 8603-8626.

25. Carlson, R. W., Johnson, R. E. and Anderson, M. S. (1999) Sulphuric acid on Europa and the radiolytic sulphur cycle, *Science* **286**, 97-99.

26. Ehrenfreund, P. and Charnley, S. B. (2000) Organic molecules in the interstellar medium, comets and meteorites, *Ann. Rev. Astron. Astrophys.* **38**, 427-483.

Glossary and Short Biographies

1. Coyne S. J., G. (2005) God's chance creation, *The Tablet* **6**, August.

2. Darwin, C. (1859) *The Origin of Species by Means of Natural Selection or the Preservation of Favored Races in the Struggle for Life*, John Murray/Penguin Books, London, 1968, p. 455. Reprinted by Oxford World's Classics, G. Beer (ed.), 1998, Oxford University Press, London.

3. Desmond, A. and Moore, J. (1991) *Darwin*, M. Joseph, London.

4. Henderson, L. J. (1913) *The Fitness of the Environment An Enquiry into the Biological Significance of the Properties of Matter*, Peter Smith, Gloucester, MA, 1970.

5. C. M. Martini (1998) *Orizzonti e limiti della scienza, Decima Cattedra di non credenti*, E. Sindoni and C. Sinigaglia (eds.). (In the series Scienze e Idee, directed by G. Giorello.) Raffaello Cortina Editore, Milano.

6. Monod, J. (1972) *Chance and Necessity an Essay on the Natural Philosophy of Modern Biology*, Collins, London.

7. Ponnamperuma, C. and Chela-Flores, J. (eds.) (1993) *Chemical Evolution: Origin of Life*, A. Deepak Publishing, Vol. **135**, Hampton, Virginia, USA.

8. Ponnamperuma, C. and Chela-Flores, J., Guest Editors (1994) *Chemical Evolution: The Structure and Model of the First Cell*, J. Biol. Phys. **120**, 1-4.

9. Chela-Flores, J. and Raulin, F. (eds.) (1996) *Chemical Evolution: Physics of the Origin and Evolution of Life* (The Cyril Ponnamperuma Memorial Conference). Kluwer, Dordrecht, The Netherlands.

10. Chela-Flores, J. and Raulin, F. (eds.) (1998) *Exobiology: Matter, Energy, and Information in the Origin and Evolution of Life in the Universe*, Kluwer, Dordrecht, The Netherlands.

11. Chela-Flores, J, Owen, T. and Raulin, F. (2001) *The First Steps of Life in the Universe*, Kluwer Academic Publishers, Dordrecht, The Netherlands.

12. Seckbach, J., Chela-Flores, J., Owen, T. and Raulin, F. (eds.) (2004) *Life in the Universe from the Miller Experiment to the Search for Life on Other Worlds*, Cellular Origin, Life in Extreme Habitats and Astrobiology, Vol. **7**.

13. Schonborn, C. (2005) Finding design in nature, New York Times, July 7.

14. Barbour, I. (1988) Five models of God and evolution, in *Evolutionary and Molecular Biology: Scientific Perspectives on Divine Action*, R. J. Russell, W. R. Stoeger S. J. and F. J. Ayala, (eds.), Vatican City State/Berkeley, California, Vatican Observatory and the Center for Theology and the Natural Sciences, pp. 419-442.

15. Russell, B. (1935) *Religion and Science*, Dover, New York, 1995.

16. Russell, B. (1991) *History of Western Philosophy and Its Connection with Political and Social Circumstances from the Earliest Times to the Present Day*, Routledge, London. p. 13.

17. McKay, C. P. (2001) The search for a second Genesis of life in our Solar System, in *The First Steps of Life in the Universe*, J. Chela-Flores, T. Owen and F. Raulin (eds.), Kluwer Academic Publishers, Dordrecht, The Netherlands, pp. 269-277.

18. Teilhard de Chardin, P. (1965) *The Phenomenon of Man*, Fontana Books, London.

19. *Ceremony in Honour of Abdus Salam on his 65th Birthday* (1991) ICTP, Trieste, p. 79.

20. Salam, Abdus (1991) The role of chirality in the origin of life, *J. Mol. Evol.* **33**, 105-113.

21. Holmes, S. J. (1961) *Louis Pasteur*, Dover, New York.

22. Weinberg, S. (2008) *Cosmology*, Oxford, Oxford University Press.

23. Weinberg, S. (1999) A designer universe? *The New York Review of Books* **46** (16), 46-48.

Abbreviations

Å: Ångstrom, unit of distance corresponding to 10^{-8} cm.

ASTEP: the Astrobiology Science and Technology for Exploring Planets program.

AU: astronomical unit (i.e., the distance Earth-Sun). This unit is explained in the Glossary, p. 171).

AUV: autonomous underwater vehicle.

Boomerang: Balloon Observations of Millimetric Extragalactic Radiation and Geophysics.

BP: Before the present.

CERN: The European Centre for Nuclear Research in Geneva, Switzerland.

COBE: Cosmic Background Explorer.

COROT: Convection Rotation and Planetary Transits mission.

CMB: cosmic "microwave" background.

DNA: deoxyribonucleic acid.

ENDURANCE: Environmentally Non-Disturbing Under-ice Robotic Antarctic Explorer.

ESA: European Space Agency.

GR: General Relativity.

Gyr BP: gigayear (10^9) years before the present (ie., a thousand million years   before the present).

HGT: horizontal gene transfer.

HR: Hertzsprung-Russell, names associated with the famous graphic display of the main-sequence stars.

HST: Hubble Space Telescope.

ICTP:	The Abdus Salam International Centre for Theoretical Physics, Trieste, Italy.
IDP:	Interplanetary dust particle.
ISAC:	Institute of Atmospheric Sciences and Climate of the Italian National Research Council (CNR).
ISS:	International Space Station.
ISAS:	the Japanese Institute of Space and Astronautical Science.
ISSOL:	The International Society for the Study of the Origin of Life/The International Astrobiology Society.

Jn:	John

LHC:	the Large Hadron Collider.
LISA:	Laser Interferometer Space Antenna.

Mic:	Micah (In the Bible, cf., The Old Testament).
MAP:	The Microwave Anisotropy Probe.
Mpc:	Million parsec.
Mt:	Matthew (In the Bible, cf., The Old Testament).
Myr BP:	million years before the present.

NASA:	National Aeronautics and Space Administration.
NGST:	Next Generation Space Telescope.
NICMOS:	Near Infrared Camera and Multi-Object Spectrometer of the HST.
NIMS:	The Galileo Near-Infrared Mapping Spectrometer.
NSF:	National Science Foundation.

PAL:	present atmospheric level.
pc:	parsec.
ppm:	parts per million.

R:	the scale factor of cosmology.
R_J:	A distance equivalent to the radius of Jupiter.
RM:	the red, aerobic, locomotor muscle.

rRNA: ribosomal RNA.
RNA: ribonucleic acid.
SETI: acronym for search for extraterrestrial intelligence.

TPF: Terrestrial Planet Finder.

UV: ultraviolet.

WIMPs: weakly interacting massive particles.
WMAP: the Wilkinson Microwave Anisotropy Probe.

yr BP: years before the present.

Glossary and Short Biographies*

Almagest

This work was *Ptolemy*'s main contribution to astronomy, whose original name had been "The Mathematical Collection". This book is divided into 13 sections, dealing with different astronomical concepts that ranged from catalogues of stars to objects of the solar system. The Almagest summarizes the knowledge of Greek astronomy; it also introduces us to the preceding work of *Hipparchus*.

Anthropic Principle in physics (the strong form)

The laws of nature and the physical constants were established so that human beings would arise in the universe.

Anthropic Principle in physics (the weak form)

Change the laws (and constants of nature) and the universe that would emerge most likely would not be compatible with life.

Anthropocentrism

This is a doctrine that maintains that man is the centre of everything, the ultimate end of nature.

Aquinas, Thomas (1225–1274)

Thomas Aquinas was of central importance. He studied under Albertus Magnus. His long association with his teacher made him a scholar deeply concerned with the Aristotelian method. He began teaching in Paris in 1257. The greatest work of Thomas was the "Summa" and the full presentation of his views. In Part I Thomas treatment of God is in much the same way as *Aristotle* had treated the *Prime Force*.

God is the first cause, himself uncaused without corporeality. He suggests a fivefold proof for the existence of God that is considered as a rational designer. God governs the world as the universal first cause.

---

* The Glossary has been provided with cross-references. Except for the terms in bold letters, all words in italics in the text itself indicate that the terms are to be found as separate entries in this Glossary. The reader is encouraged to refer to the entry that is pointed out to him, in order to grasp the subject that is being discussed.

In Part II Thomas develops his system of ethics, which has its root in Aristotle (cf., entry in this Glossary). He analyses the role of human reason in ethics. Part III is devoted to Christ and the question of incarnation. Christ as head of humanity imparts perfection and virtue to his members. He is the teacher of humanity; his whole life and suffering as well as his work after he is exalted serve this end.

Archaea (cf., archaebacteria). One of the three *domains* introduced in the *taxonomic* classification of all life on Earth by *Carl Woese.*

Archaebacteria

These are a group of single-celled organisms that are neither bacteria nor *eukaryotes.* They may be adapted to extreme conditions of temperature (up to just over 100°C), in which case they may be called thermophiles. They may also be adapted to extreme acidic conditions (acidophiles). The alternative expression of *extremophiles* is used sometimes for these organisms to distinguish different degrees of adaptability to such extreme ranges of conditions. All archaebacteria are said to form the domain *Archaea,* which is divided into kingdoms.

Archaean

In geologic time, this is an era that spans from 4.5 to 2.5 Gyr BP. (The Hadean is the first of its suberas). We refer to this era together with the *Proterozoic* (2.5-0.57 Gyr BP) as the *Precambrian Eon.* (The eon in which multicellular eukaryotic life arose is called the *Phanerozoic and* it ranges from the end of the Precambrian till the present).

Aristarchus of Samos (310–230 BC)

He was one of the last of the Ionian scientists that founded the studies of philosophy. He already formulated a complete Copernican hypothesis, well before *Nicholas Copernicus,* according to which the Earth and other planets revolve round the Sun; but in so doing, Aristarchus asserted, the Earth rotates on its axis once every 24 hours. The heliocentric theory did not prosper in antiquity. Instead, the influential astronomer *Hipparchus* (cf., entry in this Glossary), who flourished from 161 to 126 BC, adopted and developed a non-heliocentric theory ("epicycles"), which was going to dominate the ancient world right into the Middle Ages. *Ptolemy* defended the final form of the ancient model of the Solar System in the middle of the second century AD.

Aristotle (384–322 BC)

Being a Greek philosopher and scientist, Aristotle and Plato, are considered the most distinguished intellectuals of Ancient Greece. Even before the term "Renaissance Man" was introduced at the end of the Middle Ages, Aristotle deserved this recognition for the wide range of his interests, which spread over virtually the whole of human culture: from physics to chemistry, biology, zoology, botany, psychology, political theory, ethics, logic and metaphysics, history and literary theory. His philosophy supported both early Christian as well as Islamic scholastic developments.

Astrobiology

This term conveys better than *exobiology* research in the field of biological aspects of the subjects of the origin, evolution and distribution of life in the universe. Astrobiology is currently in a period of fast development due to the many space missions that are in their planning stages, or indeed already in operation. (cf., also *bioastronomy*.)

Astrometry

This subject concerns the observation of the position of celestial objects and its variation over time. Recently, astrometry has been applied to the measurable wobbling motion of a star, due to its rotating suite of planets, against a fixed background of stars. This phenomenon has been used to discover the existence of extra-solar planets.

Astronomical unit

The average distance between the Sun and the Earth is called an Astronomical unit, approximately 150 million kilometers, or equivalently 93 million miles. It is abbreviated as AU. There are about 63,240 AU in a *light year*.

Astrophysics

A branch of astronomy that is concerned with properties and structure of cosmic objects. It encompasses a variety of subjects including *nucleosynthesis* and *stellar evolution*. Cosmology and *cosmogony* can be considered to be theoretical astrophysics at the largest scales and earliest times.

Atheism

A position that rejects *theism*.

Augustine of Hippo, Saint (354–430)

He was an influential philosopher and theologian who was bishop of Hippo from 396 to 430. He unified Christian, Roman, and Platonic traditions into a philosophical system that influenced all subsequent developments of Christianity. Besides his two most influential books "Confessions" and "City of God", numerous other works—over one hundred—have survived to the present day. These writings greatly influenced biblical exegesis. From the point of view of a modern reader what is most striking in his work is the role that he assigns to reason in the unified view of philosophy and theology. For Augustine revealed knowledge can be elucidated by reason, although reason itself may be sufficient for arriving at a theological truth.

Baryons and baryonic matter

These are subnuclear particles formed by *quarks* and interact through "strong interactions", much stronger that the electromagnetic interaction.

Base

Purines or *pyramidines* in *DNA* or *RNA*.

Bergson, Henri (1859–1941)

He was a French philosopher, who was awarded the Nobel Prize for Literature in 1927. He wasthe first to elaborate what came to be called a *process philosophy*, which rejected static values in favor of values of motion, change, and evolution.

Bioastronomy

This is a synonym of *astrobiology*, but it emphasizes is more on research with the specific tools of the radio astronomers. *Frank Drake* is one of its pioneers who used extensively radio astronomy, namely the study of radio waves emitted naturally by objects in space.

Biochemistry

This is the branch of science that studies the chemistry of living organisms. (It overlaps to a certain extent with *molecular biology*).

Bioethics

This is a branch of *moral philosophy* concerned with the implications of biological research and applications. As biological research advances ethical issues arise. In astrobiology there are ethical issues concerning the contamination of other environments of the solar system during their exploration.

Biogeocentrism

A term introduced in the text to reflect a tendency observed in some contemporary scientists and philosophers according to which life is only likely to have occurred on Earth. This term is not to be confused with *Henderson*'s use of the term "biocentric".

Biogeochemistry

This is concerned with the study of life-related signatures present in the geologic record. The authenticity of a microfossil may be verified by studying the presence of *isotopes* of chemical elements that are typical of ancient living processes.

Biomarker (*or biosignature*)

This is a characteristic biochemical substance or mineral that can be taken to be a reliable indicator of the biological origin of a given sample.

Biota

This term means the totality of life on Earth.

Black hole

This is an object in space whose gravitational force is so strong that light cannot escape from its interior. In stellar evolution has been conjectured by theoretical reasoning to form when a massive star collapses at thee end of its life.

Black hole, Supermassive

These *black holes are* thought to power *quasars*. Some astronomers speculate that there may be copious numbers of black holes comprising dark matter.

Brown dwarfs

If a star's mass is less than one twentieth of our Sun, its core is not hot enough to burn either hydrogen or deuterium, so it shines only by virtue of its gravitational contraction. These dim objects, intermediate between stars and planets, are not luminous enough to be directly detectable by our telescopes. Brown dwarfs and similar objects are known collectively as massive compact halo objects (MACHOs). If dark matter is made of MACHOs, then it is likely that *baryonic matter* does make up most of the mass of the universe.

Bruno, Giordano (1548–1600)

He was a philosopher, astronomer, and mathematician, who is remembered for intuitively going beyond the heliocentric theory of Copernicus, which still maintained a finite universe with a sphere of fixed stars. From the point of view of astrobiology his anticipation of the multiplicity of worlds has been amply confirmed since 1995. In that year the first detection of extra-solar planets was announced. But what is more significant regarding Bruno's intuition is that he also conjectured that such worlds would be inhabited by living beings. His ideas went much further than those of the Englishman *Thomas Digges*, who was one of his contemporaries. Astrobiology, the science of life in the universe is just concerned with this key question, still without a convincing answer. Bruno was directly influenced by the philosophy of *Cusanus* Bruno's major innovation was his refusal to accept that the Solar System is contained in a cosmos bounded by a finite sphere of fixed stars. He suggested an infinite cosmos, populated by an infinite number of worlds. This proposal was first outlined in his first Oxford Dialogue: "The Ash Wednesday Supper". His work took place during a visit to England in 1584. These writings were in fact stimulated by controversial debates at the University of Oxford. His cosmological vision matured in Bruno's writings long before the science of *astrometry* allowed his views to be brought within the scientific domain.

Camus, Albert (1913–1960)

French novelist, essayist, and playwright, best known for such novels as "L'Étranger" (1942; "The Stranger"), "La Peste" (1947; "The Plague"), and "La Chute" (1956; The Fall). He received the 1957 Nobel Prize for Literature. He understood *nihilism* but argued in favor of defending truth and justice.

Carbonaceous chondrite

This is a meteorite formed in the early solar system. Its constituents are silicates, bound water, carbon and organic compounds, including *amino acids.*

Carbonate

This is a mineral that releases carbon dioxide by heating.

Cassini-Huygens mission

The spacecraft sent to Saturn is the most ambitious effort in planetary space exploration. A joint endeavor of ESA, NASA and the Italian space agency, Agenzia Spaziale Italiana (ASI), Cassini-Huygens is a sophisticated spacecraft being sent to the ringed planet to study the Saturnian system in detail over a four-year period. A scientific probe called Huygens was released from the main spacecraft to parachute through the atmosphere to the surface of the satellite Titan.

Chirality

The properety of existing in left- and right-handed structural forms. This property is especially important when in phenomenon of *optical activity.*

Cenancestor (cf., *progenote*).

Cenozoic

The most recent era of the Phanerozoic eon covering approximately 65 million years ago to the present.

Chondrules

These are millimeter-sized spheroidal particle present in some kinds of meteorites. Originally they were molten or partially molten droplets.

Christ

A Greek word, which is a translation of the Hebrew word *"Messiah"*. The followers of Jesus of Nazareth, led by St. Paul, used this word to describe themselves as *Christians*.

Christian

This is a name originally applied to the followers of Christ. According to the New Testament—Acts 2: 26—it was first used in Antioch about 40 AD. At present the term Christian refers to one of the religions of the world with a large number of followers.

Comte, Auguste (1798–1857)

The particular ability of this French philosopher was correlating a wide range of intellectual currents. He was under the influence of 18th and early 19th centuries writers. From *Hume* and *Kant* he derived his conception of the theory that *theology* and *metaphysics* are earlier imperfect modes of knowledge. He further assumed that positive knowledge is based on natural phenomena. His doctrine is referred to as *positivism*.

Contingency

This term denotes the notion that the world today is exclusively the result of chance events in the past.

Convergent evolution

This is an independent evolution of similar genetic or morphological features.

Copernicus, Nicholas (1473–1543)

His heliocentric theory was published posthumously in 1543. During his stay at the University of Padua from 1501 to 1503, Copernicus had been influenced by the sense of dissatisfaction of the Paduan instructors with the systems of both *Ptolemy* and *Aristotle*. The new century was a time in which scientists were requiring a sense of simplicity that classical philosophy could no longer provide. The main work of Copernicus, "De revolutionibus orbium coelestium" (On the revolution of the heavenly spheres). In this influential work Copernicus placed the Sun at the center of the Solar System and inserted both the Sun and planets inside a sphere of fixed stars. Copernicus deliberately appeared to accept the tenets of Aristotle's universe, as proposed in his "De Caelo", a cosmos inserted within a system of revolving planets, surrounded by a sphere of fixed stars. There is ample evidence that the Polish scientist knew of the heliocentric hypothesis of *Aristarchus*. Indeed, *Bertrand Russell* argues that the almost forgotten hypothesis of the Ionian philosopher did encourage Copernicus.

Cosmogony

An account of the origin of the universe. While in classical Greek philosophy *Thales of Miletus* and *Plato* made attempts, the earliest scientific accounts were developed within the gravitational theory of Sir Isaac Newton, notably by Newton himself and later by and *Emmanuel Kant*. The contemporary cosmogonical account has been tested in space by probes of the main space agencies of Europe and the United States that have successfully tested the bases of Albert Einstein's theory of gravitation (General Relativity).

Coyne S.J., George (1933–)

A member of the Society of Jesus since the age of 18. He joined the Vatican Observatory as an astronomer in 1969 and became an assistant professor at the LPL in 1970. Coyne became Director of the Vatican Observatory in 1978. He is adjunct professor in the University of Arizona Astronomy Department. Coyne's research interests include *Seyfert galaxies*, the *polarization* produced in cataclysmic variables, or interacting binary star systems that give off sudden bursts of intense energy, and dust about young stars. He has contributed significantly to the dialogue between science and faith[1].

Crick, Francis Harry Compton (1916–2004)

He was an English molecular biologist. He was one of the co-discoverers of the structure of the *DNA* molecule in 1953. He, James D. Watson and Maurice Wilkins were jointly awarded the 1962 Nobel Prize for Physiology or Medicine "for their discoveries concerning the molecular structure of *nucleic acids* and its significance for information transfer in living material".

Cusa, Nicholas of (Cusanus) (1401–1464)

The Italian cardinal Cusanus in 1440 published "De Docta Ignorantia" (On Learned Ignorance). In this book Cusanus denied the one infinite universe centered on Earth. He spoke of "a universe without circumference or center". The work of Cusanus touched on a theological question. In his system, all celestial bodies are suns representing to the same extent the explication of God's creative power. Such dialogue between a scientific question (cosmology) and theology (*Divine action*) would lead in the subsequent century, to a conjecture that anticipated the possible existence of a plurality of inhabited worlds. The person that took up this initiative was *Giordano Bruno*.

Cyanobacterium

This is a single-cell microbe without a nucleus, but capable of oxygenic photosynthesis. Some cyanobacteria appear in the fossil record in the *Archaean*. They are regarded to be ancestors of chloroplasts.

Dante Alighieri (1265–1321)

He was an Italian poet from Florence. His main work was the Divina Commedia, which is considered the greatest literary work composed in the Italian language.

Darwin, Charles Robert (1809–1882)

Charles Darwin was an English naturalist author of the theory of evolution through the mechanism of *natural selection*, (*Darwinism*). His theory, was published in "The origin of species by means of natural selection or the preservation of favored races in the struggle for life"[2]. This work led to the definitive theory of evolution that had been anticipated in earlier incomplete forms by Charles' grandfather Erasmus Darwin and independently by *Jean-Baptiste Lamarck*[3] Darwin avoided the problem of the origin of life, except from making a few speculative remarks, the so-often-quoted "warm little pond". This was a very reasonable attitude for the late 19th century, before experimental science began to address the question of the origin of life with *Fox, Haldane, Oparin, Miller, Oro, Ponnamperuma* and others.

Darwin, Erasmus (1731–1802)

The grandfather of Charles lived through the *Enlightenment*. He is best remembered for its two volumes published in the 1790s under the name of "Zoonomia, or The Laws of Organic Life". Although it was a medical treatise, it contained the formulation of the earliest speculations of biological evolution, anticipating the work of both *Lamarck* and *Charles Darwin*.

Darwinism

A theory that is due to *Darwin*. It explains the mechanism of evolution. It is based on three principles: (1) variation, (2) heredity and (3) the struggle for existence. (cf., *neo-Darwinism* and *Deterministic laws*.)

De Duve, Christian (1917–)

A Belgian cytologist and biochemist that shared the Nobel Prize for Physiology or Medicine in 1974 with Albert Claude and George Palade. De Duve discovered cell organelles. De Duve's discovery answered the question of how the powerful enzymes used by cells to digest nutrients are kept separate from other cell components. In 1947 de Duve joined the faculty of the Catholic University of Louvain, Belgium. Since 1962 he headed a research laboratory at Louvain (until his retirement in 1985) and a laboratory at Rockefeller University (until 1988).

Deism

God is not involved in the world in a personal way, but instead He is responsible for its laws. God allows the world to continue in its own way. This view simplifies problems that arise from the scientific approach to explain the universe.

Descent from a common ancestor

All variations on life are the result of the evolutionary process. All living organisms are related by descent from a common ancestor. Evolution in the science of biology is a process of descent with modification (cf., *cenancestor* and *progenote*.)

Design argument

The teleological argument is supported from an observation of the general laws of the universe, which are interpreted as revealing that the world functions toward ends or goals. This argument received its clearest expression in David Hume's analysis of the argument from design, in which the universe is seen as an orderly machine.

Deterministic laws

Deterministic laws tend to be excluded from the theory of evolution due to several reasons (i.e., the course of evolution cannot to a certain extent be predicted (cf., *Convergence*).

Digges, Thomas (1546–1595)

He was an English mathematician and astronomer contemporary of *Giordano Bruno*. He was a defender of the heliocentric theory of *Nicholas Copernicus*. But his main contribution to the question of the plurality of worlds was to depart from the Aristotelian model of fixed spheres of stars. He translated part of Copernicus's "De revolutionibus orbium coelestium" and introduced the idea of an infinite universe with the stars at varying distances in an infinite space.

Dirac, Paul (1902–1984)
An English theoretical physicist known for his work in quantum. In 1933 he shared the Nobel Prize for Physics with the physicist *Erwin Schrödinger*.

Divine action
The Mosaic traditions (Judaism, Christianity and Islam), assume a process of deliberate self-revelation of God to receptive human communities and individuals. Accompanying such forms of piety is the attribution to God of both intentions and the capacity for action as a means of expressing those intentions. Divine action might be commonly assumed in religious piety. Making sense of the concept of Divine Action is a current challenge to both philosophy and theology.

DNA
Deoxyribose nucleic acid is a substance present in every cell, bearing its hereditary characteristics.

Domain
In *taxonomy* it refers to the highest grouping of organisms, including kingdoms and lower taxons, such as phyla or divisions, orders, families, genera and species.

Drake, Frank (1929–)
He is President of the SETI Institute. In 1960 he conducted the first radio search for extraterrestrial intelligence. He is a member of the National Academy of Sciences. Drake also served as President of the Astronomical Society of the Pacific. He was a Professor of Astronomy at Cornell University (1964-1984). He served as the Director of the Arecibo Observatory.

Einstein, Albert (1879–1955)
German-American physicist who developed the special and general theories of relativity, the equivalence of mass and energy, and the photon theory of light.

Element, chemical
A substance that is irreducible to simpler substances in the sense that all its atoms have the same number of protons and electrons, but may have a different number of neutrons (cf., also "*isotope*").

Empiricism
This thesis maintains that all knowledge on matter of fact is based on experience.

Enlightenment
This was an intellectual movement of the 17th and 18th centuries in which ideas concerning God, reason, nature, and man went into a synthesis that had many supporters. Amongst the most distinguished thinkers of this period we have: Descartes, Diderot, Montesquieu, Pascal, Rousseau and Voltaire, This movement was influential on the development of art, philosophy, and politics. Reason was the main theme underlying most innovations of this period. The thinkers behind this movement searched a deeper understanding of the cosmos. Rationalists strived towards more freedom, knowledge and happiness.

Eschatology
The part of systematic theology that deals with the final destiny of the individual as well as mankind

Ethics (moral philosophy)
A branch of philosophy closely associated with other studies, such as anthropology, economics, politics, and sociology. As all philosophical pursuits it investigates the basic concepts underlying a human activity. It is concerned with what is morally good and bad with a given activity, right and wrong. Ethics is conveniently divided into: metaethics, normative ethics, and applied ethics.

Eukaryogenesis
The first appearance of the *eukaryotes*, usually believed to be partly due to the process of *symbiosis*, probably during the Late *Achaean* 2.7 billion years before the present (Gyr BP), according to current views; but certainly during the *Proterozoic*, some 1.8 Gyr BP, eukaryotes were co-existing with *prokaryotes*.

Eukaryotes
These are either single-celled, or multicellular organisms in which the genetic material is enclosed inside a double membrane, which is called nuclear envelope.

Eucarya
This term means a *domain* that contains the totality of eukaryotes, including kingdoms, such as Animalia.

Evolution (biological)
In the case of biology it is a theory that assumes various types of animals and plants to have their origin in other pre-existing types. In addition, evolutionary

theory assumes that distinguishable differences between living organisms are due to modifications that occurred in previous generations

Evolution (*cosmic*)

This theory states that the universe itself is under a continuous process of expansion. The first such theory that was called "Big Bang" by *Sir Fred Hoyle*, is based on the early 1922 theoretical model of the universe by the Russian mathematician *Alexander Friedmann*. It is also based on the 1929 observations of the American Astronomer *Edwin Hubble*. The Hubble Law states that large groups of stars or galaxies move away from each other and that the velocity of recession is proportional to their distance.

Evolution (*Lamarckian*)

An early theory of evolution, enunciated by *Jean Baptiste Lamarck* that involves the inheritance of acquired characteristics.

Existentialism

Hegel and some of his contemporaries attempted to defend faith as being logical. This movement was influenced by the growth of science since the *Enlightenment*. The need was felt for a harmonious development of human culture and the bases of natural theology had to be extended to accommodate so much new scientific knowledge. *Faith* in a Creator had to be seen in a new light. Søren Kierkegaard attacked *Hegel* in particular and the concept that religious faith was logical. Kierkegaard insisted in an opposite thesis: humans suffer a deep anxiety (and hence need religion) because one has no certainties, in fact, in more modern terms we can paraphrase the pioneers of existentialism by saying that there is no meaning to life in the universe. To put it simply life the universe is not intelligible. Friedrich Nietzsche also extended the concepts of anxiety and alienation. According to this philosopher, not only is there no logic to existence, but also a positive aspect of the human nature is to rise above the absurdity of life.

Exobiology

The study of extraterrestrial life based on astronomy, physics and chemistry, as well as the earth and life sciences (cf., *bioastronomy*).

Extremophiles

This term is explained under "*Archaebacteria*".

Faith

This concept is a subjective response to Divine truth as well as a supernatural act of the will.

Foundationalism

In epistemology the concept means that knowledge could be started from basic beliefs (which in turn support to other beliefs, understood as the "foundation", upon which all new knowledge could be inferred). Basic beliefs are assumed to be self-evident. They need not be justified by more basic beliefs.

Fox, Sidney Walter (1912–1998)

An American scientist who brought vitality and excitement to his work on the origin of life. Fox stood out as a uniquely persistent scientist. He had been making significant contributions valuable to the field of the origin, evolution and distribution of life in the universe for over half a century. The publication of his sequence analysis of amino acid residues in 1945 was to be followed by an enormous output of influential papers, which were written in collaboration with over sixty associates. The study of the origin of life has gained many insights due to his work. He was the first to synthesize a protein by heating amino acids under conditions found here on Earth. He also showed that these new thermal proteins, when placed in water, would self-organize into a primitive cell. He organized numerous conferences including the Wakulla Springs Conference in 1963. This meeting brought together both pioneers of the studies of the origin of life: *Oparin* and the English biologist *John Haldane* (1892-1964).

Friedmann, Alexander (1888–1925)

A Russian mathematician, theoretical physicist and meteorologist who found in 1922 the functional behavior of the scale factor of the universe that is usually denoted with the symbol "R". Such a (standard) model is referred to as the Friedman model. His model has been of great significance in the mathematical derivation of cosmological models from *Albert Einstein*'s general theory of relativity. Friedman was also one of the first to postulate a "Big Bang" model for the evolution of the universe, anticipating *Fred Hoyle*'s work.

Galileo Galilei (1564–1642)

An Italian physicist, astronomer, and mathematician who first contributed significantly to the development of science understood as an experimental activity. And secondly, Galileo visualized the "book of nature" as being written in mathematical terms. These contributions were two of the most significant turning points in the development of modern science. In addition, with the recently introduced telescope Galileo revolutionized astronomy by studying in detail the Jovian system. His discovery of the four largest satellites of Jupiter (Io, Europa, Ganymede and Callisto, the so called Galilean satellites) in 1610 demonstrated that there were systems in the universe that did not have the Earth at their centre, giving a strong observational support to the ideas of *Copernicus, Bruno and Digges*.

Galileo mission (1995–2003)

The mission has changed the way we look at the Solar System. The spacecraft was the first to fly past an asteroid and the first to discover a moon of an asteroid. It provided the only direct observations of a comet colliding with a planet. Galileo was the first to measure Jupiter's atmosphere with a descent probe and the first to conduct long-term observations of the Jovian system from orbit. It found evidence of subsurface saltwater on Europa, Ganymede and Callisto and revealed the intensity of volcanic activity on Io.

Genesis

This was NASA's first sample return mission sent to space since Apollo in the years 1969-1972. Genesis was launched in the year 2001 with the intention to bring back samples from the Sun itself. It was retrieved in Utah three years later, after crash-landing. Genesis collected particles of the solar wind on wafers of gold, sapphire, silicon and diamond.

Geocentric

Geocentrism is a hypothesis maintaining that the Earth is at the center of the universe. In particular, this old hypothesis maintained that the Sun was in an orbit around the Earth.

Glashow, Sheldon (1932–)

American physicist who, with *Steven Weinberg* and *Abdus Salam*, received the Nobel Prize for Physics in 1979 for their efforts in formulating the electroweak theory, which explains the unity of electromagnetism and the weak force.

Gravitational lens (or gravitational lensing)

The name derives form this effect as galaxies, for instance, appear to distort objects even further as long as they are on the same line of sight. The bending of light by a gravitational field was predicted by *Einstein*'s theory of gravitation (General Relativity). This effect was first confirmed in 1917 during an eclipse.

Hadean
This term denotes the earliest subera of the *Archaean*.

Haldane, John Burdon Sanderson **(1892–1964)**
A pioneer with *Oparin* in the studies of the origin of life on Earth. Haldane was a British geneticist, biologist, He was a very influential scientist: he studied the relationships of Mendelian genetics on evolutionary theory, enzymology and genetics, and the theoretical biology. Later on in life Haldane moved to India to conduct scientific research.

Haught, John F. **(1942–)**
He is Landegger Distinguished Professor of Theology at Georgetown University. His area of specialization is systematic theology, with a particular interest in issues pertaining to science, cosmology, ecology, and religion. He was the winner of the 2002 Owen Garrigan Award in Science and Religion and the 2004 Sophia Award for Theological Excellence. He was the chair of Georgetown's theology department between 1990 and 1995. Haught, who established the Georgetown Center for the Study of Science and Religion, is the author of several important books on the creation-evolution controversy.

Hegel, Georg Wilhelm Friedrich **(1770–1831)**
He was a German philosopher whose main contribution was to develop a dialectical scheme stressing the progress of history and the progress of ideas, firstly from thesis, secondly to antithesis and finally to a synthesis. Hegel was the leading philosopher in Germany of the post Kantian period between the 18th and 19th centuries, together with the two idealists Johann Gottlieb Fichte (1762-1814) and Friedrich Schelling (1775-1854). As an absolute Idealist inspired by Christian insights Hegel found a place for everything—logical, natural, human, and divine — in a dialectical scheme that repeatedly oscillated from thesis to antithesis and back again to a higher and richer synthesis. His influence extended to Søren *Kierkegaard*, the Marxists, the Vienna *Positivists* and G.E. Moore. One key to Hegel's philosophy lies in the central role played by the history of thought.

Heidegger, Martin **(1889–1976)**
He was a German philosopher who was influenced by existentialism. Heidegger was influenced by *Kierkegard*. His main work was published in 1927 ("Sein und Zeit").

Henderson, Lawrence Joseph **(1878–1942)**
He was an American physiologist, chemist, biologist, philosopher and sociologist. Henderson was associated with Harvard Medical School during the period

1904-1942. In "The Fitness of the Environment"[4]. He discussed the biological significance of the properties of matter and went on to develop a theory of the fitness of the environment for life.

Hipparchus (died after 127 BC)

A Greek astronomer and mathematician, who discovered the precession of the equinoxes, calculated the length of the year with certain accuracy. He is also remembered by his compilation of the first star catalogue. We know Hipparchus' work through *Ptolemy*'s astronomical compendium "*Almagest*", who was referring to a predecessor by almost three centuries.

Hoyle, Fred (1915–2001)

British astronomer a proponent of the steady-state theory of the universe: the universe is expanding and that matter is being continuously created to keep the mean density of matter in space constant. His comment on the synthesis of carbon atoms in stars led to the current interest in the *Anthropic Principles*.

Hubble, Edwin Powell (1889–1953)

American astronomer who is considered the founder of extragalactic astronomy and who provided the first evidence of the expansion of the universe.

Hume, David (1711–1776)

A Scottish philosopher, who was a recognized intellectual of the *Enlightenment*. His political essays were amongst the main influences of the constitution of the United States. He developed the empirical philosophy of Locke and Berkeley. His main philosophical work was the "Treatise of Human Nature", which was written between 1734 and 1737.

Husserl, Edmund (1859–1938)

A German philosopher who developed the doctrine of *phenomenology*. He claimed that the use of words in a meaningful way should rely on insights rather than on generalizations form experience.

Hydrothermal vent

These structures occur at the crest of oceanic ridges, producing the ascension of a very hot suspension of small particles of solid *sulfide* in water (cf., *plate tectonics*, *sea-floor spreading*).

Idealism

A name given to the philosophical theories that have the view of the external world as being created by the mind.

Interstellar dust

The ensemble of solid particles borne by the gas that occupies the space between the stars.

Isotope

One of two or more atoms of the same chemical *element* that have the same number of protons in their *nucleus*, but differ in their number of neutrons.

Jesus of Nazareth (4 BC–27 AD)

His followers called him Christ. *Christians* continued to regard Him in this way. During his public career His central and unifying theme was the imminent approach of the *Kingdom of Heaven*. He taught usually through parables recorded in the four Gospels. His impact during his lifetime was strictly restricted to the Jewish people, but later on Christianity became one of the largest religions on Earth, mainly through the preaching and the writings of St. Paul.

Kant, Immanuel (1724–1804)

A German philosopher who distinguished himself in the theory of knowledge, ethics, and aesthetics. By so doing he influenced all subsequent philosophy (often referred to as the "post-Kantian period"). Kant was the leading philosopher of the *Enlightenment* and one of the greatest philosophers of all time. In him were subsumed new trends that had begun with *Rationalism* (stressing reason) and *Empiricism* (stressing experience).

Kenosis

A term used as self-emptying and voluntary sacrifice on behalf of others, based on genuine and freely given love for others, and resulting in generosity and respect that flow from it.

Kierkegaard, Soren (1813–1855)

A Danish philosopher who wrote on religion. He emphasized the significance of the existing individual, as well as faith. His writings influenced protestant theologians, as well as the doctrine of *existentialism*.

Kingdom of Heaven

A concept typical of Hebrew thought, but extended by *Jesus of Nazareth*. In Hebrew thought God was eternally king in heaven. At the time of Jesus mission the Kingdom of Heaven would be understood in the sense of *eschatology*. Jesus originality was to emphasize the ethical and religious faith demanded of those that would aspire to the Kingdom of Heaven. Jesus seems to some extent to have implied that the Kingdom of Heaven was not present when he spoke of it. This has sometimes been interpreted in terms of Christ's *Second Coming*.

Kuiper belt

A group of some billion (10^9) comets beyond the orbit of Neptune and the source of short-period comets.

Lamarck, Jean-Baptiste (1744–1829)

At the beginning of the 19th century Lamarck in his late 50s began the revolutionary steps that led him to develop an evolutionary theory, rather than accepting that the living world was fixed and harmoniously organized. Lamarck believed that a change in the environment causes changes in the needs of organisms living in that environment, which in turn causes changes in their behavior, and once again in turn this leads to differential use of internal organs. (Eventually organs either continuously improve or are lead to their gradual disappearance.) Such a mechanism for evolution of life on Earth—called Lamarckism—was different from *Darwin*'s (particularly since Lamarck stated that these changes in the organisms would be inheritable). But, in the end, Lamarckism leads to adaptive change in lineages driven by environmental change, over geologic time. In spite of its limitations, the originality of evolution of life on Earth being driven by the environmental changes disrupted a view of life that persevered the life sciences since the time of *Aristotle*. According to Lamarck, therefore, living species are inter-related through reproduction, slowly evolving through the course of generations.

Leibniz, Gottfried Wilhelm (1646–1716)

He was born in Leipzig, the son of a Professor of Ethics (Moral Philosophy). He entered Leipzig University and eventually earned a doctorate in law. He decided against an academic career. He translated *Plato*'s Phaedo and Theaetetus into Latin, and was perhaps the first person to make a clear distinction between the doctrines of Plato, and those of *neoplatonism*. In Paris Leibniz made the acquaintance of the philosophers Arnauld and Malebranche, and the mathematician and physicist Huygens. Under Huygens, he developed the infinitesimal calculus.

While in Paris, Leibniz also worked on physics, and he wrote a treatise called The "Theory of Concrete Motion". However, his most important idea was a calculating machine, which incorporated devices that were still in use until mechanical calculators were replaced by electronic ones. At the end of 1676, Leibniz traveled to Holland, where he visited the microscopist Antonie van Leeuwenhoek, who had recently made the first observations of microorganisms; and he spent four days of intense discussion with *Spinoza*.

Light-year
The distance light travels in a year in a vacuum, approximately ten trillion miles (10^{13} Km); more precisely it is 2.99792458×10^8 m, or 63,240 *AU*.

Logical positivism
A philosophical current that maintained that scientific knowledge is the only kind of factual knowledge; all traditional doctrines are to be rejected as meaningless. Logical positivism differs from *positivism* (*David Hume*, Ernst Mach) in maintaining that the basis of knowledge depends exclusively on experimental verification instead of personal experience. It differs from *Auguste Comte* in assuming that metaphysics is not merely false, but instead it is meaningless.

Logos
A term that has had many meanings in the history of philosophy. Firstly, for Heraclitus it was some sort of non-human intelligence that has been capable of bringing order in the world. Secondly, for the Stoics this concept was stood for all the rationality in the world, essentially equivalent to some aspect of the modern idea of God. Finally, in St. John's Gospel, the Logos is equated with *Jesus of Nazareth*, the Christ, where we appreciate his redemptive aspect.

Maritain, Jacques (1882–1973)
A philosopher specialized in the writings of *Thomas Aquinas*. Maritain studied philosophy at the Sorbonne, Paris and biology at Heidelberg. He attended the lectures by the philosopher *Henri Bergson*. He insisted on the idea that moral philosophy (ethics) should take into account other branches of human knowledge.

Martini, S.J., Cardinal Carlo Maria (1927–)
Ordained in a Jesuit seminary in 1952, He graduated from the Gregorian University with a degree in Theology. In 1978 he was Rector of the Pontifical Gregorian University. On December 29, 1979 Pope John Paul II named him the new Archbishop of Milan. and was ordained by the Holy Father and installed as the

head of the important See of Milan in northern Italy. He retired in the summer of 2002. He has authoring numerous biblical, theological and spiritual thesis and documents. He directed the Chair of non-believers, where scientists, including astrobiology and evolution were able to discuss constructively the frontier of science and religion[5].

Mesozoic

An era of the *Phanerozoic eon* that was characterized by reptiles together with brachiopods, gastropods and corals. The best-known period of this era is the Jurassic period (208-146 Myr BP) with rich flora and warm climate when reptiles, mainly dinosaurs, were dominant on land.

Messenger RNA (**mRNA**)

An *RNA* molecule that specifies the *amino acid sequence* of a protein.

Messiah

A person invested by *God* with special powers and functions. The continued expectation of a deliverer is reflected in the Torah (Mic.5:2). The followers of *Jesus of Nazareth* accepted His own acknowledgement (Mt. 16:17) that He was the Messiah. King David was considered to have been specially chosen by God. The permanent expectation of a deliverer that ought to descend from the House of David is reflected in the New Testament, for example in Mt. 2:4-6.

Metaphysics

This is a philosophical subject, of central importance in Western civilization since Greek philosophy. It is concerned with the study of the real nature of things. Topics of interest to this discipline are the meaning, structure, and principles of reality as a whole. Important contributions are due to *Kant*, *Leibnitz*, and *Spinoza*. In the 20th century we may add R. G. Collingwood and P. F. Strawson.

Miller, Stanley (**1930–2007**)

He published a remarkable paper in 1953 on the generation of amino acids. It was a simple experiment that attempted to reproduce conditions similar to those in the early Earth, when life first originated. He was under the guidance of Harold Urey, who had done fundamental research in nuclear physics. Urey was responsible for the discovery of an isotope of hydrogen: deuterium. He received the Nobel Prize for this work.

Urey had subsequently suggested that the early Earth had conditions favorable for the formation of organic compounds. As a subject of his doctoral thesis Miller demonstrated experimentally that amino acids, the building blocks of the proteins could be formed without the intervention of man in environmental conditions, which we have called prebiotic—similar to those that presumably were valid at the earliest stages in the evolution of the Earth itself. The corresponding geologic period was the *Archaean*.

Miller's work was an important step in the growth of the subject of chemical evolution. The 50th anniversary of the 1953 paper was celebrated with a scientific event in Trieste, where there is further information on the relevance of this singular contribution to astrobiology[12].

Molecular biology
The study of the structure and function of the macromolecules associated with living organisms.

Monod Jacques (1910–1976)
A French molecular biologist that shared the Nobel Prize for Physiology or Medicine in 1965 for his contributions to biochemistry with François Jacob and André Lwoff. Jacob and Monod explained gene regulation of cell metabolism In 1961 they proposed the existence of messenger *RNA* (mRNA), a substance whose base sequence is complementary to that of *DNA* in the cell. They postulated that mRNA carries the information encoded in the base sequence to *ribosomes*, the sites of protein synthesis; here the base sequence of m RNA is translated into the amino acid sequence of a biological catalyst. Monod's book-length essay "Le Hasard et la nécessité"[6] argued that the origin of life is the result of chance.

Moore, George Edward (1873–1958)
A British philosopher who published in 1904 his major ethical work, "Principia Ethica". From 1925 to 1939 he was professor of philosophy there. Moore was preoccupied with such problems as the nature of sense perception and the existence of other minds and material things. He was not as skeptical as those philosophers who held that we lack sufficient data to prove that objects exist outside our own minds, but he did believe that proper philosophical proofs had not yet been devised to overcome such objections.

Moral philosophy (cf., ethics).

Natural selection
A term suggested by *Darwin* for the struggle for existence, differences in survival, in fertility, in rate of development, in reproductive success. It is a process resulting in the adaptation of an organism to its environment by means of selectively reproducing changes in its genetic constitution.

This significant contribution to science is has led to a perennial difficulty inserting it into the mainstream of cultural knowledge. We refer the reader to the dialogue between his Eminence Cardinal *Christoph Schönborn*[13] and *George Coyne* S. J.[1]

Natural theology
The body of knowledge about religion that can be obtained by human *reason* alone, without appealing to *revelation*.

Naturalism
A concept in philosophy used by *G. E. Moore* as an approach that relates scientific method to philosophy by affirming that all beings and events in the universe (whatever their inherent character may be) are natural. Consequently, all knowledge of the universe falls within the range of scientific investigation.

Neo-Darwinism
Knowledge of the mechanisms of inheritance allow us to distinguish between non-inheritable morphological variation and variation of a genuinely inheritable kind.

Neoplatonism
This doctrine consisted in the recasting and combination of Plato's philosophy with that of other Greek philosophers by the Egyptian born and Roman resident Plotinus (c.205-270 AD). Amongst the realities of the world he considered the soul and also discussed the concept of *Nous*.

Newton, Isaac (1642–1727)
English scientist who laid the foundations of modern physical optics, and formulated three laws of motion that became basic principles of modern physics and led to his theory of universal gravitation.

Nihilism
A term used by a generation of believing only in science and materialism, but went beyond positivism with their radical ideas.

Nous
Stands for knowledge and reason. *Plato* associated with it the rational part of the soul, while *Aristotle* conceived it as meaning the immortal intellect.

Nucleic acid
These are organic acids whose molecular structure consists of five-carbon sugars, a phosphate and five *bases*.

Nucleosynthesis
The generation of the chemical *elements* starting from hydrogen, which took place at an early stage of the Big Bang. It also takes place at the centre of stars and during the *supernova* stage of *stellar evolution*.

Oort cloud
A group of comets surrounding the solar system more than a thousand times more populated than the *Kuiper belt*. It is estimated to extend to 50,000 AU, or more.

Oparin, Alexander (1894–1980)
Oparin faced a problem that was basically philosophical in nature. Together with *Haldane* they laid out the basis of the studies of the origin of life.

Optical activity
A phenomenon studied in depth by *Louis Pasteur*[21]: the ability of certain substances for instance, asymmetric crystals, to rotate the plane of plane-*polarized* light as it passes through the crystal itself. The phenomenon occurs in molecules that exist in two forms whose configurations are mirror images of each other.

Oró, John/ Joan/Juan (1923–2004)
He was a Spanish biochemist whose research in the USA focused mainly on the experimental study of the chemical reactions that occurred in the primitive Earth. In 1960 he accomplished the pre-biotic synthesis of adenine from HCN, a major component of comets, and in 1961 published a theory on the role of comets on the formation of biochemical compounds that led to the appearance of life on Earth.

Paleozoic
This is the earliest era of the *Phanerozoic eon*. It is divided into the following periods: Cambrian (570-510 Myr BP) when there was a great expansion of animal, Ordovician (510-439 Myr BP), Silurian (439-408 Myr BP), Devonian (408-362 Myr BP), Carboniferous (362-290 Myr BP), and Permian (290-245 Myr BP).

Pantheism

The doctrine that identifies God with all that there is and that man and nature are not independent of God. Christian *theism* finds the identification of Nature with God unacceptably close to atheism.

Parsec

An astronomical distance equal to 3.26 light-years.

Pasteur, Louis (1822–1895)

French chemist and microbiologist who proved that microorganisms cause disease; he was the first to use vaccines for rabies, anthrax, and chicken cholera; and performed important pioneer work in *optical activity*.

Penzias, Arno (1933–)

American astrophysicist who shared the 1978 Nobel Prize for Physics with *R. Wilson* for the discovery of the cosmic microwave background (cf., Sec. 5.7).

Phanerozoic

The most recent eon of geologic time extending from the *Paleozoic* (570-230 Myr BP) to the *Mesozoic* (230-62 Myr BP) and the *Cenozoic* (since 62 Myr BP).

Phenomenology

A philosophical doctrine developed by *Edmund Husserl*. He suggested that his approach was an *a priori* search for the meanings common to the thoughts of different minds.

Philosophy of biology

The main contribution of this branch of philosophy is to separate linguistic difficulties from matters of substance.

Philosophy of nature

This is the subject concerned with investigating issues regarding the actual features of nature as a reality.

Philosophy and theology (relations)

Theism represents a constructive philosophical position not to be identified with any particular religion. Theistic lines of thought have emerged from within a given religion. For instance, Christian theism is on the boundary between philosophy and theology in its own right.

Phylogenetic tree

This is a graphic representation of the evolutionary history of a group of organisms. This may be extended from organisms to species or any higher ordering, such as kingdoms. In *molecular biology* this concept has been extended to genes as well. The main lesson we may draw form such trees is the path followed by evolution.

Phylogeny

This term means the evolutionary history of an organism, or the group to which it belongs.

Pioneers **10** *and* **11**

The Pioneer series is a long running series of space probes. Pioneer 10 was launched outward toward Jupiter bypassing it in 1973. It continued to send back data after it passed the orbit of Pluto in April 1983. Pioneer 11 was launched in 1973 and reached Saturn in 1979.

Plantinga, Alvin **(1932–)**

He is John A. O'Brien Professor of Philosophy, University of Notre Dame. Plantinga was in Ann Arbor, Michigan USA. He began to study at Harvard University in 1950, but came under the influence of William Harry Jellema at Calvin College. Plantinga returned to Calvin to study with him. Philosophy at Calvin emphasized studying the history of philosophy. He studied the history of philosophy with an emphasis on divergent religious visions.

Plato **(428/427–348/347 BC)**

A Greek philosopher, who helped significantly to the philosophy of Western civilization. Starting with the achievements of Socrates, Plato made significant contributions to logic, epistemology, and metaphysics; but his main thrust is in ethics. Essentially Plato is a rationalist, with complete confidence on reason. It has been maintained that Plato's philosophy, rests upon a basis of eternal Ideas, or Forms. His system is essentially a rationalistic *ethics*.

Pluto **(*New Horizons*)**

New Horizons will be the first spacecraft to visit the small planet Pluto (2320 km) and its suite of moons, especially Charon (1270 km) at a distance of 16 Pluto radii (R_P). The spacecraft was launched on 11 January 2006. Arrival is scheduled for July 2015. All measurements will be crammed into a short fly-by period. The final phase of the mission will be the exploration of the nearer ranges of the *Kuiper Belt*.

Polarization and polarimetry

The technique of measuring the polarization of light. When light rays exhibit different properties in different directions, the light is said to be "polarized".

Ponnamperuma, Cyril (1923–1994)

Ponnamperuma came in contact with the Nobel Laureate Melvin Calvin (1911-1997) in 1962. This led to a series of papers on the synthesis of DNA components, extending in a significant way the pioneering work initiated by *Stanley Miller*. In 1971, he joined the Maryland faculty as head of the Laboratory of Chemical Evolution, which he directed until his death. He became principal organic analysis investigator for the Apollo project and also worked on the *Viking* and *Voyager* programs. He left NASA to join the Maryland faculty. Ponnamperuma played a major role in NASA's early experiments for detecting life on Mars at the time of the Viking missions. His analysis of meteorites showed that the basic chemicals of life were not confined to the Earth. In the analysis of the meteorite that fell in Murchison, Australia in 1969, he and his co-workers provided evidence for extraterrestrial amino acids. His contact with the Nobel Laureate *Abdus Salam* led to the Trieste series of conferences in astrobiology[7-12].

Positivism

This philosophical system maintains that the goal of knowledge is simply to describe the phenomena experienced, not to question whether it exists or not. In philosophy positivism is any system that restricts itself to the data of experience to the exclusion of arguments based on metaphysics. *Auguste Comte* influenced positivism in the 19th century. He maintained that a scientific basis for reorganizing society was possible; he advocated a "positive sociology", and even suggested that positivism could be a substitute for religion.

Prime Mover

The Prime Mover was a concept introduced in the Aristotelian cosmology that was capable of triggering the initial, eternal rotation at constant angular velocity of the outermost crystalline sphere. This motion propagated from sphere to sphere, hence inducing the whole cosmos to rotate.

Process philosophy (*or process thought*)

Process philosophy holds that what exists in nature is not just originated and sustained by processes, but in fact they characterize it.

Process theology

An approach to natural theology based on the metaphysics of *Alfred North Whitehead*, who rejects Divine action in terms of causality, proposing that God acts persuasively in all events, but not necessarily determining their character. God's power over events in the world is severely limited, especially regarding events that are mostly determined by their past.

Process theology and Christianity
Christian process theologians maintain that the life and death of Christ are examples of the love of God and participation in the life of the world[14].

Progenote or cenancestor
These are terms introduced by the American evolutionist Carl Woese to denote the earliest common ancestor of all living organisms (cf., descent from a common ancestor).

Prokaryotes
These are unicellular organisms that lack a nuclear envelope around their genetic material. Normally they are smaller than nucleated cells (*eukaryotes*). Well-known examples are bacteria. All prokaryotes are encompassed in two *domains*: Bacteria and *Archaea*.

Protein
An organic compound, which is an essential biomolecule of all living organisms. Its elements are: hydrogen, carbon, oxygen, nitrogen and sulfur. It is made up of a series of *amino acids*. (A medium-size protein may contain 600 amino acids.)

Proterozoic
In geologic time this is an era that ranges form the *Archaean* era (4.5-2.5 Gyr BP) to the *Phanerozoic Eon* (570 Myr BP to the present time).

Providence
This concept is rooted in the belief in the existence of a benevolent, wise, and powerful deity or a number of beings that are benevolent and that are either fully divine or, at least, appreciably wiser and more powerful than man.

Ptolemy (active in Alexandria from AD 127–145)
He was the most influential astronomer of the first millennium who considered the Earth at the centre of the universe. The Ptolemaic system is introduced in the first section of the *Almagest*. Ptolemy rather complex explanation of the retrograde motion of planets with the Earth immovable at the centre of the cosmos prevailed in Western culture for over a millennium. We had to wait for the contributions of *Copernicus*, *Digges* and *Bruno* to have the present-day view of the cosmos.

Purines and pyrmidines

These are two groups of nitrogenous bases making up part of both deoxyribonucleotides (*DNA*) and ribonucleotides (*RNA*), and the basis for the universal genetic code.

Quarks

These are constituents of subnuclear particles (e.g., the proton) that interact through a short-range force, much stronger that the electromagnetic interaction. Their existence was suggested by symmetries and detected by collisions.

Raphael Sanzio (**1483–1520**)

A great master of the Renaissance. A large body of his work remains in the Vatican, including the *Stanza della Segnatura*.

Rationalism

The doctrines that believe in the possibility that knowledge of nature can be obtained by reason alone.

Reason

A faculty contrasted with experience, passion or *faith*.

Redemption

A concept that occurs in several religions. In *natural theology*, redemption is understood in terms of deliverance from sin, as well as the restoration of mankind to communion with *God*.

Reductionism

The ideas that physical bodies are collections of atoms or that thoughts are combinations of sense impressions are forms of reductionism. Two general forms of reductionism have been held by philosophers in the 20th century:

(1) *Logical positivism* has maintained that expressions referring to existing things or to states of affairs are definable in terms of directly observable objects, or sense-data, and, hence, that any statement of fact is equivalent to some set of empirically verifiable statements. In particular, it has been held that the theoretical entities of science are definable in terms of observable physical things, so that scientific laws are equivalent to combinations of observation reports. (2) Proponents of the unity of science have held the position that the theoretical entities of particular sciences,

such as biology or psychology, are definable in terms of those of some more basic science, such as physics; or that the laws of these sciences can be explained in terms of more basic science. The logical positivist version of reductionism also implies the unity of science insofar as the definability of the theoretical in terms of the observable would constitute the common basis of all scientific laws.

Regolith

The general term regolith applies to unconsolidated residual or transported material that overlies the solid rock on the earth, moon, or a planet. The case that concerns us is the loose, fragmental material on the Moon's surface. It is the product of meteoritic bombardment, due to the debris thrown out of the impact craters that form the uppermost surface of the Moon. The composition and texture of the lunar regolith varies according to the rock types that collided with the Moon.

Revelation

This is a process by which communication of truth by God takes place. Christian philosophers have distinguished between "truths of reason" and "truths of revelation" (cf., *Natural theology*.) Revelation the monotheistic religions state that God has chosen to manifest himself through the prophets, but He can manifest Himself through his main creations the universe and the life that has evolved in it. In the Judeo-Christian tradition the prophets are witnesses and interpreters of God's *Divine action*, both in their transmission of God's messages, as well as in the way to interpret His Divine action.

RNA

Ribonucleic acid is a substance present in all organisms in three main forms: *messenger RNA*, ribosomal RNA, and transfer RNA (used in the last stage of translation during protein synthesis). Their general function is related with the translation of the genetic message from *DNA* to *proteins*.

Russell, Bertrand Arthur William (1872–1970)

Russell was an English logician and philosopher, whose seminal work in mathematical logic was published in the early 20th century. Russell collaborated with *Alfred North Whitehead* on "Principia Mathematica" (1910-1913). He received the Nobel Prize for Literature in 1950. His interests ranged from philosophy and mathematics to religion. His excellent books on the history of philosophy, as well as his book on science and religion, have been used repeatedly in the present work[15,16].

Sagan, Carl Edward (1934–1996)

An American astronomer who searched for intelligent life in the cosmos. Sagan began researching the origins of life in the 1950s and went on to play a leading role in every major U.S. spacecraft expedition to the planets.

Salam Abdus (1926–1996)

He made many contributions to the creation of the theory that unifies the electromagnetic and weak nuclear interactions, sharing the 1979 Nobel Prize in Physics. (It was shared with *Sheldon Glashow* and *Steven Weinberg*. Salam encouraged *Ponnamperuma* and the present author to initiate a series of conferences on astrobiology, a series that spread over a period of 11 years. He was the Founding Director of the International Centre for Theoretical Physics (ICTP) from 1964. Salam remained in that position till the end of his life. Salam would state in 1991[19]: "*I feel particularly proud of my last paper*[20] *on the role of chirality on the origin of life.*"

Sartre, Jean-Paul (1905–1980)

A French philosopher, who together with *Heidegger* became leaders of *existentialism*. He published his main philosophical work in 1943 (Etre e le Néant). As a novelist he also expressed concern with the nature of human existence.

Schoenborn, Cardinal Christoph (1945–)

An Austrian cardinal who was appointed as Vienna's archbishop in 1995. He has renewed the dialogue between science and faith[13]. *Coyne* has challenged his theological views[1].

Schrödinger, Erwin (1887–1961)

An Austrian theoretical physicist who contributed to the fundamentals of quantum mechanics. He shared the 1933 Nobel Prize for Physics with *Dirac*.

Second Coming

The assumption that there will be a future return of Christ to judge the living and the dead. A precise prediction of when this event may occur has never been explicitly stated (cf., *Kingdom of Heaven*).

Second Genesis

This is a term used as a synonym for the emergence of life in the universe away from planet Earth. Christopher McKay introduced this expression in the context used in this book[17].

Seyfert galaxies
A group of spiral galaxies with very small and unusually bright star like centers.

Solar wind
This phenomenon is the expansion of the solar corona to form a flux of protons, electrons and nuclei of heavier elements that are accelerated by the high temperatures of the solar corona, or outer region of the Sun, to high velocities that allow them to escape from the Sun's gravitational field. The solar wind is so tenuous that at a distance of 1 *AU* during a relatively quiet period, the wind contains approximately five particles per cubic centimeter moving outward from the Sun at velocities of 3×10^5 to 1×10^6 ms^{-1}.

Space Weather
Interplanetary space is filled mainly with high-energy particles and radiation, forming clouds of hot plasma emitted by the Sun. Space weather studies how the Earth environment is influenced by the Sun.

Spinoza, Baruch (1632–1677)
He was a Dutch philosopher. In 1656 he was expelled from the synagogue for his opinions. His philosophical treatises earned him a reputation in Europe. His "Theologico-Political Treatise" of 1670 is an examination of popular religion and a critique of the Protestantism that was practiced in Holland. Spinoza criticized anthropomorphic conceptions of God. He suggested modern methods for biblical exegesis, and favored toleration of religious practices. Speculations on *metaphysics* dominated Spinoza's philosophy. He attempted to formulate a rationalistic view of the universe as a unitary whole. At the end of 1676, the scientist and philosopher *Leibniz* traveled to Holland, where he spent four days of intense discussions with Spinoza, whose influential book "Ethics" was published the following year. This is a fundamental work on moral philosophy.

Stanza della Segnatura (*Room of the Signatura*)
This was the first room to be decorated by *Raphael Sanzio* in the Palace of the Vatican. (The study housing the library of Julius II, in which the Signatura of grace tribunal was originally located.) The spirits of Antiquity and Christianity are brought into harmony reflecting the contents of the pope's library including theology, philosophy and science, for instance astronomy—the Prime Mover ceiling panel (1509-1511).

Stellar evolution

Stars constantly change since their condensation from the interstellar medium. Eventually they exhaust their nuclear fuel and the *supernova* stage may occur. At this stage *elements* synthesized in its interior are expelled, enriching the interstellar medium, out of which new generations of stars are be born.

Stromatolite

A geological feature consisting of a stratified rock formation that is essentially the fossil remains of bacterial mats (mainly *cyanobacteria*). Similar mat-building communities can develop analogous structures of in the world today.

Supernova

The late stage in *stellar evolution*, in which an evolved massive star has exhausted its nuclear fuel. This leads to a catastrophic stellar explosion. It enriches the *interstellar dust*. Its remnant is a *neutron star*, which can be detected as a pulsar. (These are sources of regular burst of radio waves.)

Taxonomy, taxonomic classification

This is the study of the theory, practice and rules of classification of living or extinct organisms into groups, according to a given set of relationships.

Teilhard de Chardin, Pierre (1881–1955)

He was a French philosopher and also a paleontologist. His best-known work is a theory of evolution of man toward a spiritual unity with *God*. His work appeals to an overlap of science and *natural theology*. He was ordained a priest in 1911. Most of Teilhard's scientific output was concerned with mammalian paleontology. His main work on philosophy, already mentioned in the text[18], was published in 1955. Teilhard's combined approach included theology, science and philosophy. He attempted to show that what is of permanent value in traditional philosophical thought can be maintained and even integrated with a modern scientific outlook if one accepts that the tendencies of material things are directed, either wholly or in part, toward the production of higher, more complex beings. Teilhard regarded basic trends in physics as being ordered toward the production of progressively more complex entities of atoms, molecules, cells and organisms, until the human body evolved, with a nervous system sufficiently sophisticated to permit rational reflection, self-awareness, and moral responsibility.

Teilhard turned away from an exclusively scientific, when he accepted to convert his work into *natural theology*. Indeed, Teilhard suggested that when humanity and the world have reached their final state of evolution, the *Second*

Coming of *Christ* would initiate a new convergence between them and God. Teilhard asserted that the work of *Jesus of Nazareth* is to lead the material world to this cosmic *redemption*.

Teleology

Teleology may be considered as either a doctrine according to which everything in the cosmos has been designed with Man in mind. Alternatively it can be interpreted as a theory of phenomena that is to be explained in terms of its purpose, rather than by initial causes. *Kant* contributed significantly on teleological judgments, especially in his "Critique of Judgment".

Terrestrial planets

These are the inner rocky planets (Mercury, Venus, Earth, Mars). Due to the similar characteristics of the Moon it is convenient to include it in this group.

Thales of Miletus (flourished in the 6th century BC)

He was the first of Greeks to enquire into what we understand today as philosophy. Thales attempted to use reason rather than myth to explain nature. In other words, Thales searched for causes in natural phenomena rather than in the gods of the Greek pantheon.

Aristotle regarded Thales as the founder of Western philosophy. Thales was the first thinker to put himself the question of what was the nature of the to substratum of the universe. He suggested that it was water.

Theism

The belief in *God* that is a central element in the Judeo-Christian tradition views God as caringly related to the world. It has tended to affirm both the immanence and the transcendence of this ultimate being, the creator of everything that exists. Doctrines of God are independent of the traditions found in religion.

Theology

A branch of the humanities that is concerned with the study of religious thought. The science of the Divinely revealed religious truths (*revelation*). Its main themes of interest are: *God,* man, the world, salvation, and a study (sometimes referred to as *eschatology*) of the final events in the history of mankind.

Theology (Kenotic Process Theology)

An approach to natural theology emphasizing God as the sole ground for the world's being. It attempts to explain the world in terms of evolution as understood within the Darwinian tradition. This view have been defended by *Haught*.

Theology (*Process Theology*)

An approach to natural theology based on the metaphysics of *Whitehead* who rejects *Divine action* in terms of causality, proposing that God acts persuasively in all events, but not necessarily determining their character.

Theology (*Natural theology*)

The body of knowledge about religion that can be obtained by human *reason* alone, without appealing to *revelation*.

Ulysses

This is the name given to an ESA-NASA solar probe launched in 1990 that made the first-ever measurements of the Sun in six-year orbits over the Sun's poles. In route to the solar polar orbit, it made a significant close flyby of Jupiter.

Viking missions

This was the first mission to land a spacecraft safely on the surface of Mars and return images of the surface. NASA built two identical spacecrafts, each consisting of a lander and an orbiter. The Viking 1 lander touched down on Chryse Planitia, while the Viking 2 lander settled down at Utopia Planitia. The two landers conducted three biology experiments designed to look for possible *biomarkers*. These experiments provided no clear evidence for the presence of living microorganisms in soil near the landing sites. The combination of solar UV that saturates the surface, the extreme dryness of the soil and the oxidizing nature of the soil chemistry prevent the formation of living organisms in the Martian soil. Viking Orbiter 1 continued for four years, concluding its mission August 7, 1980, while Viking Orbiter 2 functioned until July 25, 1978. Viking Lander 1 made its final transmission to Earth in 1982. Data from Viking Lander 2 arrived at Earth two years earlier.

Voyagers **1** *and* **2**

The Voyagers are two US spacecrafts that explored the outer Solar System during the period 1977-1989. Although the two Voyagers were planned to depart the same year, they were programmed to explore different aspects of the Solar System. Voyager 2 completed a set of images of every planet of the Solar System. Close-by images of the "dwarf" planet *Pluto* will be completed by NASA in 2015. Because Voyager 2 crossed the *solar wind* termination shock, about 16 billion kilometers away from Voyager 1 and almost 1.6 billion kilometers closer to the sun, it confirmed that our solar system is "squashed" or "dented"—that the bubble carved into interstellar space by the solar wind is not perfectly round.

Weinberg, Steven **(1933–)**
American nuclear physicist who in 1979 shared the Nobel Prize for Physics with *Sheldon Lee Glashow* and *Abdus Salam* for work in formulating the theory, that explains the unity of electromagnetism with the weak nuclear force. The American Philosophical Society awarded him the Benjamin Franklin Medal, with a citation that said he is "considered by many to be the preeminent theoretical physicist alive in the world today". Besides his scientific research, Steven Weinberg has been a prominent public spokesman for science, writing articles for The New York Review of Books. His books on science have been influential[22], written for the public combine the popularization of science with philosophy of science and *atheism*[23].

Whitehead, Alfred North **(1861–1947)**
He was an English mathematician, logician and philosopher, who collaborated with *Bertrand Russell* in the writing of "Principia Mathematica" (1910-1913). From the mid-1920s, Whitehead taught at Harvard University. While he was working at Harvard from 1924 onwards, he developed a comprehensive theory of metaphysics, which has been called *process philosophy*.

Wilson, Edward Osborne **(1929–)**
He is an American biologist. By applying the evolutionary principles attempted to explain the behavior of the social insects in order to understand the social behavior of animals, including humans. These were the bases of Wilson socio-biology,a new scientific field. In his 1998 book "Consilience: The Unity of Knowledge", Wilson discusses methods that have been used to unite the sciences, and might be able to unite the sciences with the humanities.

Wilson, Robert Woodrow **(1936–)**
American radio astronomer who shared, with *Penzias*, the 1978 Nobel Prize for Physics for a discovery that supported the big-bang model of creation.

Wittgenstein, Ludwig (1889–1951)

He is one of the most influential philosophers of the 20th century. His early work was influenced by that of Arthur Schopenhauer and by his teacher *Bertrand Russell*. This work culminated in the "Tractatus Logico-Philosophicus", his only book on published during his lifetime. "Philosophical Investigations" was published shortly after he died at the age of 62 years. His works were influential in analytic philosophy.

Woese, Carl (1928–)

He is an American microbiologist who defined the Archaea as a new *domain* in the *taxonomic* tree. By 1990 Woese adopted the term "domain" for the three new branches of life: Bacteria, *Archaea*, and Eucarya. He shortened the name *Archaebacteria* to Archaea. A three-domain system is based upon genetic relationships rather than morphological similarities (cf., Chapter 6, Ref. 9).

WMAP (Wilkinson Microwave Anisotropy Probe)

By making accurate measurements of fluctuations in the *CMB* since 2001 (cf., Sec., 5.7, p. 50), WMAP is able to measure the basic parameters of the Big Bang model (cf., Sec. 5.6, p. 50), including the density and composition of the universe. If our ideas about the origin and evolution of galaxies and large-scale structure are correct, then WMAP is measuring the density of *baryonic* and non-baryonic matter to an accuracy of better than 5%. It is also able to determine some of the properties of the non-baryonic matter: the interactions of the non-baryonic matter with itself, its mass and its interactions with ordinary matter all affect the details of the CMB fluctuation spectrum.

Science

HISTORY OF SCIENCE

Bertola, F. (1992) *Da Galileo alle Stelle*, Biblos Edizioni, Padua. [This book was published during the celebrations at the University of Padua to commemorate the fourth century of the nomination of Galileo Galilei as a lecturer of Mathematics in that university, where he spent 18 years of very productive work.]

Bertola, F. (1995) *Imago Mundi, la rappresentazione del cosmo attraverso i secoli*, Biblos, Cittadella PD, Italy.

Bertola, F. (2001) The Plurality of Worlds, In *The First Steps of Life in the Universe*, J. Chela-Flores, T. Owen and F. Raulin (eds.), Kluwer Academic Publishers, Dordrecht, The Netherlands, pp. 401-407.

Desmond, A. and Moore, J. (1991) *Darwin*, M. Joseph, London.

Biblioteca Casanatense (2000) Giordano Bruno 1548-1600. Mostra Storico Documentaria (Roma 7 Giugno-30 Settembre). L. S. Olschki Editore. Biblioteca di Bibliografia Italiana 164: Citta' di Castello (2000).

Ricci, S. (2000) *Giordano Bruno nell'Europa del Cinquecento*, Salerno Editrice, Roma.

Moorhead, A. (1971) *Darwin and the Beagle*, Penguin, London.

ASTROBIOLOGY (THE ORIGIN OF LIFE IN THE UNIVERSE)

Cairns-Smith, A. G. (1985) *Seven Clues to the Origin of Life*, Cambridge University Press, London.

Cosmovici, C. B., Bowyer, S. and Werthimer, D. (eds.) (1997) *Astronomical and Biochemical Origins and the Search for Life in the Universe*, Editrice Compositore, Bologna.

Davies, P. (1998) *The Fifth Miracle The Search for the Origin of Life*, Allan Lane The Penguin Press, London.

Delsemme, A. (1998) *Our Cosmic Origins From the Big Bang to the Emergence of Life and Intelligence*, Cambridge University Press, London.

Greenberg, J. M., Mendoza-Gomez, C. X. and V. Pirronello (eds.) (1993) *The Chemistry of Life's Origins*, Kluwer Academic Publishers, Dordrecht, The Netherlands.

Impey, C. (2007) *The Living Cosmos*: *Our Search for Life in the universe*, Random House, New York.

Oro, J., Miller, S. L. and Lazcano, A. (1990) The origin and early evolution of life on Earth, *Ann. Rev. Earth Planet. Sci.* **18**, 317-356.

Schopf, J. W. (1999) *Cradle of Life*, Princeton University Press, Princeton.

Schopf, J. W. (ed.) (2002) *Life's Origin The Beginnings of Biological Evolution*, University of California Press, Berkeley.

Shapiro, R. (1994) *L'origine de la vie*, Flammarion, Paris.

ASTROBIOLOGY (THE EVOLUTION OF LIFE IN THE UNIVERSE)

Colombo, R., Giorello, G. and Sindoni, E. (1999) *Origine della Vita Intelligente nell'Universo* (*Origin of Intelligent Life in the Universe*), Edizioni New Press, Como, Italy.

Coyne S. J., G., Giorello, G. and Sindone, E. (eds.) (1997) *La Favola dell'Universo*. Piemme, Como.

De Duve, C. (1995) *Vital Dust. Life as a Cosmic Imperative*, Basic Books, A Division of Harper-Collins Publishers, New York.

De Duve, C. (2002) *Life Evolving Molecules Mind and Meaning,* Oxford University Press, New York.

ASTROBIOLOGY (THE DISTRIBUTION OF LIFE IN THE UNIVERSE)

Billingham, J., Heyns, R., Milne, D., Doyle, S., Klein, M., Heilbron, J., Ashkenazi, M., Michaud, M., Lutz, J. and Shostak, S. (eds.) (1999*) Social Implications of the Detection of an Extraterrestrial Civilization*: *A Report of the Workshops on the Cultural Aspects of SETI*, SETI Press, Mountain View, California, USA.

Carle, G. C., Schwartz, D. E. and Huntington, J. L. (eds.) (1992) *Exobiology in Solar System Exploration*. NASA SP 512.

Chadha, M. S. and Phondke, B. (1994) *Life in the Universe,* Publications and Information Directorate, New Delhi, India.

Crick, F. (1981) *Life Itself*, MacDonald & Co, London.

Davies, P. C. W. (1995) *Are We Alone*: *Implications of the Discovery of Extraterrestrial Life*, Penguin Books, London.

Drake, F. and Sobel, D. (1992) *Is There Anyone Out There*? *The Scientific Search for Extraterrestrial Intelligence*, Delacorte Press, New York, pp. 45-64.

Jakosky. B. (1998) *The Search for Life in Other Planets*, Cambridge University Press, New York.

Sullivan III, W. T. and Baross, J. A. (2007) *Planets and Life*, Cambridge University Press, p. 604.

Tough, A. (ed.) (2000) *When SETI Succeeds*: *The Impact of High Information Content*, Foundation for the Future, Washington, USA.

ASTROBIOLOGY (THE DESTINY OF LIFE IN THE UNIVERSE)

Dick, S. J. (1998) *Life on Other Worlds*, Cambridge University Press, London.

Dick, S. J. (2004) The new universe, destiny of life, and cultural implications, in *Life in the Universe: From the Miller Experiment to the Search for Life on Other Worlds*, J. Seckbach, J. Chela-Flores, T. Owen and F. Raulin (eds.), Kluwer Academic Publishers, Dordrecht, pp. 319-326.

TRIESTE SERIES ON CHEMICAL EVOLUTION AND THE ORIGIN OF LIFE

Ponnamperuma, C. and Chela-Flores, J. (eds.) (1993) *Chemical Evolution: Origin of Life*, A. Deepak Publishing, Hampton, Virginia, USA.

Chela-Flores, J., Chadha, M., Negron-Mendoza, A. and Oshima, T. (eds.) (1995) *Chemical Evolution: Self-Organization of the Macromolecules of Life.* A. Deepak Publishing: Hampton, Virginia, USA (1995).

Ponnamperuma, C. and Chela-Flores, J. (eds.) (1995) *Chemical Evolution: The Structure and Model of the First Cell*, Kluwer Academic Publishers, Dordrecht, The Netherlands.

Chela-Flores, J. and Raulin, F. (eds.) (1996) *Physics of the Origin and Evolution of Life*, A Cyril Ponnamperuma Memorial, Kluwer Academic Publishers, Dordrecht, The Netherlands.

Chela-Flores, J. and Raulin, F. (eds.) (1998) *Exobiology. Matter, Energy, and Information in the Origin and Evolution of Life in the Universe*, Kluwer Academic Publishers, Dordrecht, The Netherlands.

Chela-Flores, J., Owen, T. and Raulin, F. (eds.) (2001) *The First Steps of Life in the Universe*, Kluwer Academic Publishers, Dordrecht, The Netherlands.

Seckbach, J., Chela-Flores, J., Owen, T. and Raulin, F. (eds.) (2004) *Life in the Universe: From the Miller Experiment to the Search for Life on Other Worlds*, Kluwer Academic Publishers, Dordrecht, The Netherlands.

CELLULAR ORIGIN AND LIFE IN EXTREME HABITATS AND ASTROBIOLOGY

Seckbach, J. (ed.) (1999) *Enigmatic Microorganisms and Life in Extreme Environmental Habitats*, Series on Cellular Origin and Life in Extreme Habitats, (COLE) Vol. 1, Kluwer Academic Publishers, Dordrecht, The Netherlands.

Seckbach, J. (ed.) (2000) *Journey to Diverse Microbial Worlds: Adaptation to Exotic Environments*, Cellular Origin and Life in Extreme Habitats, (COLE) Vol. 2, Kluwer Academic Publishers, Dordrecht, The Netherlands.

Seckbach, J. (ed.) (2004) *Origins, Genesis, Evolution and the Origin of Life,* Cellular Origins, Life in Extreme Habitats and Astrobiology (COLE), Vol. 6, Kluwer Academic Publishers, Dordrecht, The Netherlands.

Seckbach, J. (ed.) (2006) *Life as We Know It,* Cellular Origins, Life in Extreme Habitats and Astrobiology (COLE),Vol. 11, Springer, Dordrecht, The Netherlands.

GENERAL REVIEWS
(In English)

Chela-Flores, J. (2001) *The New Science of Astrobiology from Genesis of the Living Cell to Evolution of Intelligent Behavior in the Universe,* Kluwer Academic Publishers, Dordrecht, The Netherlands.

Chela-Flores, J., Lemarchand, G. A. and Oro, J. (2000) *Astrobiology from the Big Bang to Civilisation,* Kluwer Academic Publishers, Dordrecht, The Netherlands.

Ward, P. D. and Brownlee, D. (2000) *Rare Earth. Why Life is Uncommon in the Universe,* Copernicus, Springer-Verlag, New York.

Cairns-Smith, A. G. (1985) *Seven Clues to the Origin of Life,* Cambridge University Press, Cambridge (UK).

Dyson, F. (1985) *Origins of life,* Cambride University Press, Cambridge (UK).

(In French)

Brack, A. and Raulin, F. (1991) *L'evolution chemique et les origines de la vie,* Masson, Paris.

Raulin, F. (1994) *La vie dans le cosmos,* Flammarion, Paris.

Heidmann, J. (1996) *Intelligences extra-terrestres,* Editions Odile Jacob, Paris.

Raulin, F., Raulin-Cerceu, F. and Schneider. J. (1997) *La Bioastronomie,* Presses Universitaires de France, Paris, p. 128.

(In Italian)

Sindoni, E. (1997) *Esistono gli Extraterrestri?* Il Saggiatore, Milan, Italy, p. 123.

Messerotti, M. (1990) *Ipotesi sull'esistenza di civilita' extraterrestri nello spazio*, Societa' Editoriale Libraria, Trieste, Italy.

EARTH SCIENCES

Konhauser, K. (2007) *Introduction to Geomicrobiology*, Blackwell Publishing, Malden, MA.

Rollinson, H. R. (2007) *Early Earth Systems: A Geochemical Approach*, Blackwell Publishing, Malden, MA.

SPACE WEATHER

Moldwin, M. B. (2008) *An Introduction to Space Weather*, Cambridge University Press, Cambridge.

Lilensten, J. (ed.) (2007) *Space Weather. Research Towards Applications in Europe*, Astrophysics and Space Science Library (ASSL) Series, Vol. 344, Springer, Dordrecht, The Netherlands.

LIFE SCIENCES OTHER THAN ASTROBIOLOGY

Attenborough, D. (1981) *Life on Earth*, Fontana, London.

Bendall, G. S. (ed.) (1983) *Evolution from Molecules to Men*, Cambridge University Press, London.

Bertola, F., Calvani, M. and Curi U. (eds.) (1994) *Origini: l'universo, la vita, l'intelligenza*, Il Poligrafo, Padua, Italy.

Cela Conde, C. J. and Ayala, F. J. (2007) *Human Evolution: Trails from the Past*, Oxford University Press, Oxford.

Conway Morris, S. (1998) *The Crucible of Creation. The Burgess Shale and the Rise of Animals*, Oxford University Press, London.

Conway Morris, S. (2003) *Life's Solution Inevitable Humans in a Lonely Universe*, Cambridge University Press.

Darwin, C. (1859) *The Origin of Species by Means of Natural Selection or the Preservation of Favoured Races in the Struggle for Life*. John Murray, London. Reprinted: G. Beer (ed.) (1998) Oxford World's Classics, Oxford University Press.

Gould, S. J. (1991) *Wonderful Life the Burgess Shale and the Nature of History*, Penguin Books, London.

Henderson, L. J. (1913) *The Fitness of the Environment an Enquiry into the Biological Significance of the Properties of Matter*, Peter Smith, Gloucester, Mass., 1970.

MacDonald, I. R. and Fisher, C. (1996) *Life without light*, National Geographic Magazine, October, pp. 87-97.

Margulis, L. and Sagan, D (1987) *Microcosm*, Allen & Unwin, London.

Maynard Smith, J. (ed.) (1982) *Evolution Now*, Nature, London.

Maynard Smith, J. (1993) *The Theory of Evolution*, Cambridge University Press, London.

Mayr, E. (1991) *One Long Argument Charles Darwin and the Genesis of Modern Evolutionary Thought*, Penguin Books, London.

Patterson, C. (1978) *Evolution*, British Museum (Natural History), London.

Rizzotti, M. (2000) *Early Evolution from the Appearance of the First Cell to the First Modern Organisms*, Birkhauser Verlag, Basel.

Schrödinger, E. (1967) *What is Life?* Cambridge University Press, London.

Wilson, E. O. (1992) *The Diversity of Life*, Penguin Books, London.

PHYSICAL SCIENCES

Barrow, J. D. (2007) *New Theories of Everything*, Oxford, Oxford University Press.

Barrow, J. D. and Tipler, F. J. (1986) *The Anthropic Cosmological Principle*, Oxford University Press.

Greene, B. (2005) *The Fabric of the Cosmos*: *Space, Time, and the Texture of Reality*, Knopf, New York, p. 569.

Quinn, H. R. and Nir, Y. (2008) *The Mystery of the Missing Antimatter*, Princeton University Press, Princeton, N. J.

Sagan, C. (1980) *Cosmos*, Random House, New York.

Weinberg, S. (1977) *The First Three Minutes*, Fontana/Collins, London.

Weinberg, S. (1993) *Dreams of a Final Theory,* Vintage, London.

Philosophy

HISTORY OF PHILOSOPHY

Plato (1974) *The Republic*, Penguin Classics, London.

Ayer, A. J. (1982) *Philosophy in the Twentieth Century*, Unwin Paperbacks, London.

Monod, J. (1972) *Chance and Necessity An Essay on the Natural Philosophy of Modern Biology*, Collins, London.

O'Grady, P. (2002) *Relativism*, Acumen Publishing Limited, Chesham (Buks), U.K.

PHILOSOPHY, SCIENCE AND THEOLOGY

Ambrosi, E. (2005) *Il bello del relativismo*, Marsilio Editore, Venezia.

Ayer, A. J. (1991) *The Central Questions of Philosophy*, Penguin, London.

Barrow, J. D., Conway Morris, S., Freeland, S. J. and Harper, C. L. (eds.) (2008) *Fitness of the Cosmos for Life*: *Biochemistry and Fine-Tuning*, Cambridge University Press.

Changeux, J.-P. and Ricoeur, P. (2000) *What Makes us Think?* Princeton University Press, Princeton, N. J.

Colombo, R., Giorello, G., and Sindoni, E. (eds.) (1999) *Origine della Vita Intelligente nell'Universo* (*Origin of Intelligent Life in the Universe*). Edizioni New Press, Como.

Dennett, D. C. (1995) *Darwin's Dangerous Idea*, Penguin, London.

Impey, C. and Petrry, C. (eds.) (2002) *Science and Theology*: *Ruminations on the Cosmos*, Vatican Observatory and Templeton Foundation, Philadelphia, USA.

Eligio, P. Rigamonti, G. and Sindoni, E. (eds.) (1997*) Scienza, Filosofia e Teologia di Fronte alla Nascita dell'Universo* (*Reflections on the Birth of the Universe: Science, Philosophy and Theology*), New Press, Como.

Russell, B. (1991) *History of Western Philosophy and Its Connection with Political and Social Circumstances from the Earliest Times to the Present Day*, Routledge, London.

Russell, B. (1998) *The Problems of Philosophy*, Oxford University Press.

Russell, R. J., Stoeger S. J., W. R. and Coyne S. J., G. (eds.) (1995) *Physics, Philosophy and Theology. A Common Quest for Understanding*, 2nd edition, Vatican Observatory Foundation, Vatican City State.

Wilson, E. O. (1998) *Consilience the Unity of Knowledge*, Alfred A. Knopf, New York.

THE PHILOSOPHY OF BIOLOGY

Dawkins, R. (1989) *The Selfish Gene*, Oxford University Press.

Dawkins, R. (1983) *The Extended Phenotype*, Oxford University Press.

Dawkins, R. (1988) *The Blind Watchmaker*, Penguin Books: London.

Rizzotti, M. (ed.) (1996) *Defining Life. The Central Problem of Theoretical Biology*, University of Padua, Padua, Italy.

ETHICS

Aristotle (1998) *Nicomachean Ethics*, Dover, New York.

Singer, P. (1999) *Practical Ethics*, 2nd edition, Cambridge University Press, Cambridge.

Spinoza, B. (2002) *Ethics*, Everyman, London.

Singer, P. (ed.) (1994) *Ethics*, Oxford Readers, Oxford University Press.

PHILOSOPHICAL QUESTIONS RAISED BY ASTROBIOLOGY

Dear, P. (2006) *The Intelligibility of Nature: How Science Makes Sense of the World*, The University of Chicago Press, Chicago, USA.

Rees, M. (2003) *Our Final Century*, Heinemann, London.

Zuckerman, B. and Hart, M. H. (1995) *Extraterrestrials. Where are they?*, 2nd edition, Cambridge University Press.

ESSAYS

Medawar, P. (1996) *The Strange Case of the Spotted Mouse and Other Classic Essays on Science*, Oxford University Press.

Theology

THE DIALOGUE BETWEEN FAITH AND REASON

Collins, F. S. (2006) *The Language of God a Scientist Presents Evidence for Belief*, Simon & Schuster, New York.

Davies, P. C. W. (1992) *The Mind of God*, Penguin, London.

Haught, J. F. (2001) *Theology after contact religion and extraterrestrial intelligent life*, Ann. *New York Acad. Sci.* **950**, 296-308.

Gingerich, O. (2006) *God's Universe*, Harvard University Press.

Greenspan, L. and Anderson, S. (eds.) (1999) *Russell on Religion*, Routledge, London.

Martini, C. M. (1999) *X Cattedra dei non credenti*; *Orizzonti e limiti della scienza*, Raffaello Cortina Editore, Milan, Italy.

Polkinghorne, J. (1996) *Scientists as Theologians*, SPCK, London.

Ruse, M. (2000) *Can a Darwinian Be a Christian? The Relationship Between Science and Religion*, Cambridge University Press, New York.

Russell, B. (1935) *Religion and Science*, Dover, New York.

SELECTED WORKS IN THEOLOGY

Chardin, P. T. de. (1965) *The Phenomenon of Man,* Fontana Books, London.

Martini, C. M. (2006) *Il mio novecento*, Centro Ambrosiano, Milan, Italy.

St. Augustine (1984) *City of God*, Penguin Classics, London.

St. Augustine (1961) *Confessions*, Penguin Classics, London.

SERIES ON "SCIENTIFIC PERSPECTIVES ON DIVINE ACTION"

Russell, R. J., Stoeger S. J., W. R. and Ayala, F. J. (eds.) (1998) *Evolutionary and Molecular Biology*: *Scientific Perspectives on Divine Action*, Vatican Observatory and the Center for Theology and the Natural Sciences, Vatican City State/Berkeley, California, USA.

Russell, R. J., Murphy, N. and Isham, C. J. (1996) *Quantum Cosmology and the Laws of Nature. Scientific Perspectives on Divine Action*, 2nd edition, Vatican Observatory Foundation, Vatican City State.

Russell, R. J., Murphy, N. and Peacocke, A. R. (eds.) (1995) *Chaos and Complexity. Scientific Perspectives on Divine Action*, Vatican Observatory Foundation, Vatican City State.

Index

The Author

Professor Julian Chela-Flores was born in Caracas, República Bolivariana de Venezuela. He studied in the University of London, England, where he obtained a Ph.D. in quantum mechanics (1969). He was a researcher at the Venezuelan Institute for Scientific Research (IVIC) from 1971 till 1978 and Professor at Simón Bolívar University (USB), Caracas until his retirement in 1990. During his USB tenure he was Dean of Research from 1979 till 1985. He is a Fellow of The Latin American Academy of Sciences (Caracas), The Academy of Sciences of the Developing World (Trieste), the Academy of Creative Endeavors (Moscow) and a Corresponding Member of the Venezuelan Academia de Fisica Matematicas y Ciencias Naturales (Caracas). His current positions are Staff Associate of the Abdus Salam International Center for Theoretical Physics (ICTP), Trieste, Research Associate, Dublin Institute for Advanced Studies (DIAS) and Professor-Titular, Institute of Advanced Studies (IDEA), Caracas. His area of expertise is astrobiology, in which he is the author of numerous papers including some articles published in the frontier between astrobiology and the humanities (philosophy and theology). He organized a series of Conferences on Chemical Evolution and the Origin of Life from 1992 till 2003. All the proceedings of the series of conferences have been published in American and European publishing houses. He is also the author of the book: The New Science of Astrobiology From Genesis of the Living Cell to Evolution of Intelligent Behavior in the Universe, published in 2001 and reprinted as a paperback in 2004.